Geology on your Doorstep

Geology on your Doorstep

The role of urban geology in earth heritage conservation

Edited by

Matthew R. Bennett, Peter Doyle
School of Earth Sciences, University of Greenwich, UK

Jonathan G. Larwood & Colin D. Prosser
English Nature, Peterborough, UK

1996
Published by The Geological Society

THE GEOLOGICAL SOCIETY

The Society was founded in 1807 as The Geological Society of London and is the oldest geological society in the world. It received its Royal Charter in 1825 for the purpose of 'investigating the mineral structure of the Earth'. The Society is Britain's national society for geology with a membership of around 8000. It has countrywide coverage and approximately 1000 members reside overseas. The Society is responsible for all aspects of the geological sciences including professional matters. The Society has its own publishing house, which produces the Society's international journals, books and maps, and which acts as the European distributor for publications of the American Association of Petroleum Geologists, SEPM and the Geological Society of America.

Fellowship is open to those holding a recognized honours degree in geology or cognate subject and who have at least two years' relevant postgraduate experience, or who have not less than six years' relevant experience in geology or a cognate subject. A Fellow who has not less than five years' relevant postgraduate experience in the practice of geology may apply for validation and, subject to approval, may be able to use the designatory letters C Geol (Chartered Geologist).

Further information about the Society is available from the Membership Manager, The Geological Society, Burlington House, Piccadilly, London W1V 0JU, UK. The Society is a Registered Charity, No. 210161.

Published by The Geological Society from:
The Geological Society Publishing House
Unit 7, Brassmill Enterprise Centre
Brassmill Lane
Bath BA1 3JN
UK
(*Orders*: Tel. 01225 445046
Fax 01225 442836)

First published 1996

British Library Cataloguing in Publication Data
A catalogue record for this book is available from the British Library.

ISBN 1-897799-54-3

Distributors

USA
AAPG Bookstore
PO Box 979
Tulsa
OK 74101-0979
USA
(*Orders*: Tel. (918) 584-2555 Fax (918) 560-2652)

Australia
Australian Mineral Foundation
63 Conyngham Street
Glenside
South Australia 5065
Australia
(*Orders*: Tel. (08) 379-0444 Fax (08) 379-4634)

India
Affiliated East-West Press PVT Ltd
G-1/16 Ansari Road
New Delhi 110 002
India
(*Orders:* Tel. (11) 327-9113 Fax (11) 326-0538)

Japan
Kanda Book Trading Co.
Tanikawa Building
3-2 Kanda Surugadai
Chiyoda-Ku
Tokyo 101
Japan
(*Orders*: Tel. (03) 3255-3497 Fax (03) 3255-3495)

Printed by City Print (Milton Keynes) Ltd, Bletchley, Milton Keynes MK3 7QT, UK.

Contents

Urban geology and civil engineering

Part Three: Awareness and use of the urban geological resource

The role of the local authority

Urban geology and education

Increasing public awareness and involvement

Part Four: Creating an urban geological resource

List of contributors

George M. A. Barker
English Nature, Northminster House, Peterborough PE1 1UA, UK

Matthew R. Bennett
School of Earth Sciences, University of Greenwich, Medway Towns Campus, Pembroke, Chatham Maritime ME4 4AW, UK

Greg Carson
The Wildlife Trusts, The Green, Witham Park, Waterside South, Lincoln LN5 7JR, UK

Timothy J. Charsley
British Geological Survey, Sir Kingsley Dunham Centre, Keyworth, Nottingham NG12 5GG, UK

Alan Cutler
Black Country Geological Society, 21 Primrose Hill, Wordsley, Stourbridge DY8 5AG, UK

Jane Dove
School of Education, University of Exeter, Heavitree Road, Exeter EX1 7RU, UK

Peter Doyle
School of Earth Sciences, University of Greenwich, Medway Towns Campus, Pembroke, Chatham Maritime ME4 4AW, UK

George R. Fenwick
School of the Environment, University of Sunderland, Benedict Building, St George's Way, Sunderland SR2 7BW, UK

Anna Grayson
The Science Unit, BBC, Broadcasting House, London W1A 1AA, UK

Mike Harley
English Nature, Northminster House, Peterborough PE1 1UA, UK

Duncan Hawley
Earth Science Teachers' Association, The Geological Society, Burlington House, Piccadilly, London W1V 0JU, UK

Thomas A. Hose
Faculty of Leisure & Tourism, Buckinghamshire College, Queen Alexandra Road, High Wycombe HP11 2JZ, UK

Susan M. Ingham
School of Earth Sciences, University of Greenwich, Medway Towns Campus, Pembroke, Chatham Maritime ME4 4AW, UK

Ed A. Jarzembowski
Brighton Borough Council, Bartholomew Square, Brighton BN1 1JA, UK

Simon Knell
Dept of Museum Studies, University of Leicester, 105 Princess Road East, Leicester LE1 7LG, UK

Jonathan G. Larwood
English Nature, Northminster House, Peterborough PE1 1UA, UK

Mike McGibbon
School of Earth Sciences, University of Greenwich, Medway Towns Campus, Pembroke, Chatham Maritime ME4 4AW, UK

Steven G. McLean
Tyne & Wear Museums, The Hancock Museum, Barras Bridge, Newcastle-upon-Tyne NE2 4PT, UK

Roger Mason
Centre for Extra-mural Studies, Birkbeck College, 26 Russell Square, London WC1B 5DQ, UK

Kevin N. Page
English Nature, Northminster House, Peterborough PE1 1UA, UK

Eileen J. Pounder
Faculty of the Built Environment, University of the West of England, Frenchay Campus, Coldharbour Lane, Bristol BS16 1QY, UK

Colin D. Prosser
English Nature, Northminster House, Peterborough PE1 1UA, UK

Colin Reid
Dudley Museum & Art Gallery, St James's Road, Dudley DY1 1HU, UK

Eric Robinson
Geologists' Association Librarian, Dept of Geological Sciences, University College London, Gower Street, London WC1E 6BT, UK

Graham J. Worton
38 Vale Road, Netherton, Dudley DY2 9HZ, UK

Preface

The rocks and landforms of Britain contain a unique story; a story of mountain building, climate change and the evolution of life. Only by conserving these rocks and landforms can we tell this amazing story.

Conservation is about education and raising public awareness. To be successfully conserved, geology must be valued not just by a handful of geologists, but by all. Over 80% of the population in the United Kingdom live in urban areas: it is to these people that we must take the message of geology if we are to conserve our earth heritage. The geological potential of urban areas has long been documented and described. However, its conservation and use in raising public awareness of geology is a relatively new initiative. With this in mind, a one-day conference was organized by the University of Greenwich and English Nature in order to explore urban geology and its role in conservation. This volume is the product of that conference and illustrates not only the breadth of the urban geological resource but its potential in education, for raising public awareness and for earth heritage conservation as a whole.

The book is divided into four parts. In the first part, a series of chapters explores the importance of urban geology and the scope of the urban geological resource. In the second part, the nature of the geological resource of urban areas is considered. The third part of the volume contains chapters which consider the need for public awareness of urban geology. The final section comprises a chapter with practical suggestions of how we can improve the nature of our urban geological resource. It is hoped that the chapters within this volume will stimulate interest in urban geology and increase awareness of its importance in raising the public profile of the earth sciences.

Matthew R. Bennett, Peter Doyle,
Jonathan G. Larwood & Colin D. Prosser
July 1995

Acknowledgements

The editors thank all those who have helped shape their ideas on urban geology, in particular Eric Robinson who has championed the cause of urban geology for over a decade and repeatedly demonstrated its importance to earth science conservation as a whole. The editors also thank all those who helped organize the one-day conference at the University of Greenwich in January 1995 on which this volume is based, in particular: Jane Anderton, Fiona Cocks, Tony Cosgrove, Martin Gay, Angela Holder, Adrienne Meredith, Alice Philpot, Richard Pole, John Roberts and Lisa Templeman. This volume has benefited from constructive reviews by Christopher Green, Mick Stanley and Chris Wilson. The production of this volume has been made possible by Pat Brown, Martin Gay, Louise Goodwin, and Nicola Hunn. The final assembly of this book was facilitated by the contribution of our research assistant, Angela Holder.

Part One

The rationale and scope of earth heritage conservation in urban areas

Part One demonstrates the scope of the urban geological resource and its role in wider earth heritage conservation. Within rural areas, site conservation is the most important focus for earth heritage conservation. Sites - quarries, pits, natural outcrops and landforms - are rare in the urban environment. Buildings, graveyards and parks are not. This part contain the chapters which reflect the changing nature of the urban geological resource, the intimate relationship of urban development with geology, and the need to use urban geology in promoting a greater awareness of earth heritage conservation.

1 The rationale for earth heritage conservation and the role of urban geology

Matthew R. Bennett & Peter Doyle

Summary

- The nature, scope and need for earth heritage conservation in Britain is reviewed.
- Increased public awareness is identified as being crucial for effective conservation.
- Since the majority live in urban areas it is in this environment that there is the highest priority for increasing the public awareness of geology.
- Urban geology is an under-used resource in the promotion of Earth heritage conservation issues; greater awareness of geology in the urban environment will lead to effective conservation.

The aims of this chapter are to set out a rationale for earth heritage conservation and to illustrate the essential role that urban geology has to play within it. This chapter establishes the aims of urban geology in the context of earth heritage conservation and provides a broad introduction to the aims of this volume.

Geology and geomorphology are the twin sciences of the Earth. Together, as earth science, they embrace the study of the formation, construction and development of our planet since its birth, 4600 million years ago. The rocks and landforms present at the Earth's surface provide the clues from which the Earth's history and the processes which have shaped it can be deduced. The rocks and landforms of Britain contain a unique story; a story of mountain building, climate change and the evolution of life; it is only by conserving them that we can tell this story.

Britain has an unequalled earth heritage. Modern earth science was born here. For generations, geologists and geomorphologists have studied Britain's rocks and landforms. Within its small geographical area, Britain has an unrivalled record of 2900 million years of earth history. The rocks and landforms at its surface form the landscape we see today, a landscape which makes Britain a distinct and pleasant land.

Britain's geology and landscape is of profound importance to its environment and people. The distribution of habitats, plants and animals is dependent upon the geology which underlies and forms the landscape. The economic well-being of Britain, now as in the past, is also underpinned by its geology. Prior to the Industrial Revolution the pattern of agriculture, forestry and settlement was determined by the rocks, soils and landforms of Britain's landscape. Since then the economic exploitation of mineral reserves - coal, oil and gas - has secured Britain's position as a major industrial nation. The building and character of our urban areas reflect Britain's geology and its regional variation. Geology is part of the architectural heritage of Britain. On a cultural level, earth science has also had a profound influence on human society. Poets, landscape artists and musicians have all been inspired by the beauty of the landscape. Britain's earth heritage is a rich one.

Britain is, however, a crowded island. Its natural environment is under constant threat, while its urban environment is in a constant cycle of development and redevelopment. Britain's earth heritage is a finite resource, a resource both irreplaceable and in danger. Protecting it is important to us all.

The rationale for earth heritage conservation

A small proportion of Britain's natural geological and geomorphological heritage has statutory protection under the 1981 Wildlife and Countryside Act. The standard rationale for the protection of earth science Sites of Special Scientific Interest is based upon seven major concepts which underpin the basis for their conservation.

- The well-being of the natural world is inextricably linked to our own welfare and its resources underpin every aspect of our lives. The character and diversity of geology and geomorphology, as expressed in the Earth's natural features, are therefore essential to human well-being and as such need to be conserved.
- The need to conserve a resource for the training and education of future Earth scientists, at all levels, from the youngest child to the most senior student.
- The need to conserve a resource for both present and future scientific research. Continuing scientific discovery means that models and ideas

are constantly evolving and advancing. Many of today's unresolved geological problems may be solved in the future through the application of new techniques or ideas. This will only be possible if a resource of sites and localities is conserved for future study.

- The need to conserve reference sites of international standing. Many of the chronostratigraphical units of the standard geological time-scale used throughout the world were named in or after areas of Britain. For example, the Cambrian, Ordovician and Silurian systems all stem from Latin names given to Welsh tribes in Roman Britain, and are based on rocks which outcrop in Wales and in the Welsh Borderland. The Devonian System, the Ludlow Series and the Oxfordian and Bathonian stages, all take their names from well known English places. The stratigraphical records of many of these areas serve as international references (stratotypes) of relevance to geologists all over the world. Other reference sites include those where rocks, minerals and fossils were first described. For example, Scotland's characteristic Celtic terms have provided names for a wealth of minerals and rock types, including lanarkite, benmoreite, mugearite and harrisite.
- Britain's geological heritage is of historical importance. Many milestones of geological thought have emerged from the study of British localities. These historic sites need to be conserved.
- Rocks and landforms determine the character of our landscape and have controlled, in part, the distribution of different habitats.
- British geology has had a profound effect upon landuse, the exploitation of economic minerals, and on the location and architectural character of our urban areas. Geology is part of the fabric of our social and historical culture.

Earth heritage conservation: the role of urban geology

The natural geological resource in Britain is rich and diverse and comprises coastal cliffs, upland exposures, quarries, cuttings and landforms of both the upland and lowland landscape. The scientific, aesthetic and cultural aspects of this resource are formally protected in Britain via a two-part system. (1) A network of nationally important sites designated as Sites of Special Scientific Interest (SSSI) by the conservation agencies, English Nature, the Countryside Commission for Wales and Scottish Natural Heritage. These sites are afforded legal protection under the Wildlife and Countryside Act (1981). SSSI are designated solely on scientific grounds. (2) A network of regionally important sites, county sites, designated as Regionally Important Geological/Geomorphological Sites (RIGS) by county-based voluntary RIGS groups. These sites have no legal status but are protected through the co-operation of the local planning authority, and through inclusion in local structure plans. RIGS are selected on scientific, aesthetic, cultural and educational grounds.

To be successful, however, in not only conserving designated sites but in the wider conservation of our natural heritage, the single most important goal is to raise public awareness and interest in geology and its conservation. Conservation is about education and public awareness within the disinterested majority, not just within the geological community. If we are to conserve our geological/geomorphological heritage successfully, we need to educate the general public and thereby exercise control through popular opinion on the planners, developers and decision-makers who have the potential to threaten this heritage. To be conserved, geology must be valued not simply by a few elite scientists, but by all. By raising the public's awareness of geology we can further all aspects of earth science conservation, and to do this effectively we need to bring people into contact with geology. Since over 80% of the population in the United Kingdom live in urban areas, the role of urban geology is of paramount importance. Discussion of this resource and its use is one of the essential aims of this volume.

For the majority, however, geology is a forgotten subject. This has not always been the case; in Victorian times geology was one of the most exciting and popular of all the sciences. It was an integral part of the Victorian passion for natural history and was fashionable with all classes of society. That great edifice to the Victorian ideals in arts, technology and science, the Albert Memorial in Kensington, has four statues dedicated to the 'greater sciences': astronomy, geometry, chemistry and geology. Today, geology is no longer a source of excitement. Geology is taken for granted. The cars in which we drive, the materials which make up our houses, all of these are derived from geology (Fig. 1.1). The urban

Fig. 1.1. This DeSoto car illustrates how geology is all around us. Geological materials provide: the iron and chrome out of which the body is built; the sand out of which the glass is made; the oil from which the plastics are obtained; and crude oil, of course, provides the fuel which makes it go.

environment contains the biggest reminders of the value of geology to the modern world (Fig. 1.2), yet it is in the urban environment where there is least evidence of its appreciation. We need to rekindle the interest in geology within the urban majority. As a single objective, increasing awareness has the potential for a far greater impact on earth heritage conservation than any legal system of site protection.

Urban geology as explored in this volume is not about conservation *per se* but about the role which the urban environment can play in raising public awareness of geology, so that earth science can be appreciated, whatever its location. The urban environment is so rich in geology that it provides a perfect opportunity to promote Britain's earth heritage and the best place to increase awareness of the value of geology to our everyday lives.

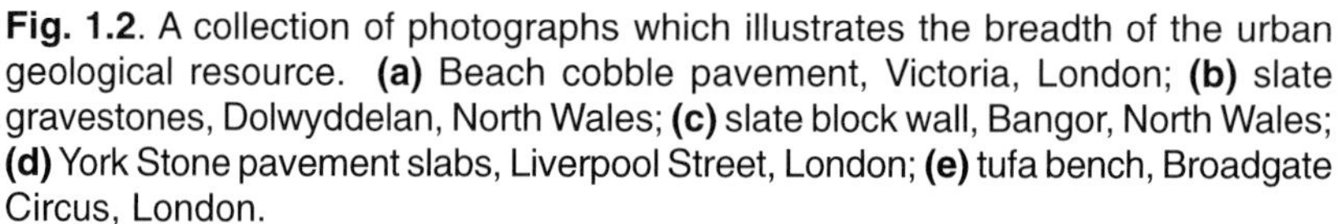

Fig. 1.2. A collection of photographs which illustrates the breadth of the urban geological resource. **(a)** Beach cobble pavement, Victoria, London; **(b)** slate gravestones, Dolwyddelan, North Wales; **(c)** slate block wall, Bangor, North Wales; **(d)** York Stone pavement slabs, Liverpool Street, London; **(e)** tufa bench, Broadgate Circus, London.

Conclusion: a rationale for urban geology

Urban geology is vital to all those interested in earth heritage conservation. It provides perhaps the greatest resource with which to raise public interest in earth science and its conservation. The rationale for the promotion of urban geology as part of any earth heritage conservation strategy can be summarized with reference to five main points.

- Eighty per cent of the population in the United Kingdom lives in urban areas. Only a small proportion of the population comes into regular contact with the geology of our natural landscape, yet most make use of it every day.
- To survive, geology needs to have a constant influx of enthusiasts and new recruits. This is essential to the life blood of the subject. Geology is now part of the National Curriculum and there is a vital need for a geological resource for the urban school pupil. After all, it is these children who will wield the power to conserve or destroy our natural earth heritage in the future.
- Conservation is about raising awareness and winning the hearts and minds of the general public - the message needs to have a realism and immediacy to be successful. A heritage value placed on the local environment will be more effective than trying to raise concern about some remote site which most people have never visited.
- Urban geology has a greater potential for participant involvement; an essential aim is to raise the profile and interest in conservation.
- As a nation, Britain is proud of its history and architectural heritage. This heritage is only possible because of its geology, yet this is not routinely mentioned or discussed.

These five points provide the agenda for urban earth science awareness and conservation. This book and its papers are intended to promote the urban earth sciences, through reaching the target audience of geologists, conservationists and planners who are all in a position to promote urban geology. In recent years the conservation agenda in Britain has been dominated by the need to recognize, designate and document a network of both nationally (SSSI) and regionally (RIGS) important earth heritage sites. It is now time to turn our attention to the wider issue of raising public awareness of geology and its conservation. Urban geology has an essential role to play in this process. The aims of urban geology as part of earth heritage conservation are listed below:

- to raise the profile of the earth sciences in public perception;
- to seek community involvement in geology and conservation;
- to seek acknowledgement for the role of geology within the urban or architectural landscape and within the social and economic infrastructure of our society;
- to conserve the urban geological exhibits of previous generations;
- to conserve the small geological resources which exist in urban areas;
- to provide a teaching resource for education within urban areas.

These aims provide the potential to fulfil the wider aim of reestablishing geology as one of the most important of our sciences: a science which has the greatest potential for community involvement and one in which there is a pressing need for greater awareness of its vulnerability and for conservation.

2 Urban geology: mapping it out

Timothy J. Charsley

Summary

- Geology has been of great importance to the economic development and success of certain urban areas.
- The geological heritage of such urban areas is therefore intimately linked with its historical development.
- The geology of many towns still influences their development today.
- The British Geological Survey provides a variety of services which cater for the continuing need for geological information to support urban development.

It is common for textbooks describing the development of urban Britain to dwell on the various geographical reasons why urban centres are situated where they are. The dry ground in the meander bend, the tidal limits on a river, the hill commanding a through valley, the list is endless. The books are, of course, right; most conurbations in Britain have historical roots going back to the time when security for defenders, and reliable water and food supplies were the reasons for the presence of the more populous settlements. This is, however, only half the story. It says why they were sited where they were but not why the population remained there and grew over the centuries.

Britain is full of lost towns and villages which have been discovered by ploughmen, historians and archaeologists during the last hundred years or so. It is also full of towns and villages which have population numbers now similar to those in the Middle Ages, but equally has many that have grown substantially. The key to economic success and longevity lies beyond geography and commonly resides in geology. The reasons for original settlement may well have been geographical, but for historical continuity, geology will have been the most important factor.

The serious student of urban area development should therefore look closely at geology, starting with the geological map, to elucidate the historical and cultural reasons for urban growth. The map will show the deposits underlying the area and which of these have or had economic value to the inhabitants. It will also show what potential there was for augmenting the water supply, and will show the limits to expansion in terms of the proximity of floodplains, soft ground, unstable slopes or other potential natural hazards.

A specific example of this applies in Britain. Survival of major population centres through the Industrial Revolution depended largely on the proximity of outcropping workable coals. It was the wealth or employment created by the working of the coal, or the uses to which it could be put, often with other mineral raw materials, that created growth (Prosser & Larwood 1994). Some wool towns of Domesday Book pedigree such as Loughborough in the Midlands, which was able to feed off the economic benefits of the Derbyshire Coalfield, continued to prosper into the eighteenth century, whereas time stood still (perhaps very much to our aesthetic advantage) for the similarly sized, but geologically rather barren, Lavenham in Suffolk.

Geology of some British cities

Edinburgh is an example of an extremely successful city, remaining prosperous over the last thousand years. Geology can indicate why this is so. Table 2.1 summarizes key periods in Edinburgh's history and the associated important geological factors of each time (McAdam 1994). Throughout its history, various aspects of the local geology have supported the growth of the city and, although the last 20 years have seen a decline in the local mineral industry, including coal production, there has been an increase in quarrying of hard rock aggregates for construction. Perhaps it can be said that Edinburgh has come full circle. Mineral production has returned to the local, hard volcanic rocks from which the city was born.

Table 2.1. Edinburgh, the growth of the city and geology (after McAdam 1994).

Date	Event
2-3000BC	Volcanic hills provide dry places and stone rubble for strongholds. Springs at the junction of basalts and sedimentary rocks provided a water supply
c. 80AD	Roman fort and port at Cramond, with small dolerite island lying just offshore
12th century	Old Town constructed from Old Red Sandstone, with quarries in red stone at Bruntsfield and Craigmillar in close proximity to town
1780s	New Town constructed from white and pale grey sandstone from the Carboniferous Oil Shale Group with quarries at Craigleth and Hailes near the growing town. Coal and limestone extraction was also very active
From 1850s	Increasing production of Oil Shale and Coal, with both reaching a production peak in the 1920s
Last 20 years	Hard rock aggregates won from volcanic rocks, as from the quarry on Torphin Hill

Nottingham provides another example of a city which has remained prosperous from at least Saxon times onwards. What role has the geology played in all this? When William the Conqueror arrived there were two settlements on the north bank of the Trent, Snotingaham and Snotingaton - the first on Sherwood Sandstone and the second on Mercia Mudstone. For William it was no contest when he came to make his decision as to where to build a castle in 1068; he chose the Sherwood Sandstone. Fortunately, the Normans had a slight pronunciation problem and the inhabitants were saved from living in Snottingham, named after the Anglo-Saxon ruler Snot.

William had decided on the site for sound geological reasons, as shown in Table 2.2. The sandstone held water but was dry and free draining above the water table, and provided solid building foundations. It was also easily carved and Nottingham became renowned as the City of Caves. The outcropping coals in Coal Measures and building stone from the Magnesian Limestone just to the west of the city, added to the Triassic brick clays and gypsum to the east and south, and sands and gravels along the Trent, ensured that the city thrived through the Middle Ages and Industrial Revolution.

Table 2.2. Nottingham, the city and geology (after Charsley *et al.* 1990).

Factors relevant to the success of Nottingham as a city through the ages:
River navigable to sea - early bridge crossing possible
Dry sandstone site - well above frequent winter floods
Sherwood Sandstone - a major aquifer easily tapped by wells
Caves readily created in sandstone beneath houses to give storage space, especially for beer and wine
Sited between major exposed and concealed coalfields
Local supply of brick clay, building stone, gypsum, sand and gravel

National geological mapping programmes

Following the establishment of the Geological Survey in 1835, successive governments supported a national programme to produce reliable, up-to-date geological maps (Wilson 1985). Initially the maps showed only the solid bedrock, but the familiar solid and drift maps followed as the programme evolved from small-scale reconnaissance to detailed larger scale surveys. The programme was given impetus at successive periods by differing national requirements. During Victorian times, it was hastened by major construction projects, such as those for the canals and railways. During World Wars I and II, the need was for general mineral resource evaluation. The survey of the coalfields became the main drive to produce large-scale maps, particularly between the wars and post-World War II following the nationalization of the coal industry.

The present phase of resurvey, where it applies to British cities and other urban areas, has largely been supported as part of the Department of the Environment's Applied Geological Mapping Programme. BGS has prepared reports and applied geology maps for 47 urban areas, including some in all parts of the United Kingdom. A selected list of these areas, including some in Scotland, Wales and from the north to the south of England, is given in Table 2.3.

Applied geology or thematic maps are designed to provide a broad coverage of environmentally significant geoscientific factors tailored to local requirements. They are specifically directed by the Department of the Environment to support policies for guidance and legislation relating to urban planning and development. A progressive programme has been in operation since 1982. Typical components making up a set of urban applied geological maps are listed in Table 2.4. This whole development

Table 2.3. Selected list of areas covered by BGS urban applied geology mapping, supported by the Department of the Environment.

Bournemouth-Poole	Bridgend
Black Country	Coventry
SE Edinburgh	Glasgow
SE Leeds	Nottingham
Southampton	Stoke on Trent
Note: including the above, a total of 47 reports have been published on geological aspects of land use planning and development in urban areas.	

Table 2.4. Typical themes covered by BGS applied geology maps in addition to Solid and Drift geology.

Thickness of superficial deposits
Distribution of man-made and worked ground
Geomorphology, drainage and slope stability
Mineral resources, mining and quarrying
Hydrogeology
Engineering geology
Geological factors for consideration in land use planning

into broad aspects of geoscience represents a considerable enhancement of basic geological information into new datasets, which are mostly now held as digital databases.

The LOCUS project

London is the subject of a new BGS initiative indicative of the way geological mapping is going. The BGS London Computerised Underground and Surface Geology Project (LOCUS for short) represents the new approach to digitizing geological information to produce customized products. The geology of London has obviously not changed since the six inches to one mile surveys of the 1890s, 1910s and 1930s, but the information available has increased enormously over the last few decades as engineering geological consultancies have become more methodical and professional in their recording procedures. Collection and databasing, particularly of borehole records, have enabled BGS not only to make the 2D mapped geological boundaries more accurate, but also to produce digital 3D models from which maps and sections of the subsurface can be derived. This allows contour and isopachyte (equal thickness) maps of the main subsurface layers to be produced to order; a major advance in the information available for planning and design of engineering construction, particularly for the utilities and transport routes under Central London.

Future trends

For the future, databasing and digitization to produce 3D models of subsurface geology will continue, with the emphasis on those urban areas where there is most demand from industry and where costs may be more readily recovered from multiple users. The continued development of map- and non-map-based systems will allow the further integration of databases to include geology, geochemistry, geophysics and hydrogeology, but increasingly there will be moves to extend the range of databases through use of Geographical Information Systems (GIS) to incorporate ecological, atmospheric, medical, sociological and other factors. The first BGS project to attempt this is in progress for the city of Wolverhampton, where the methodology for such an ambitious multidisciplinary approach is being researched. The initial results are very encouraging.

Digital map production is now routine in the BGS and this not only allows customized output as hard copy or in digital formats, but gives complete freedom from scale. The digital geological map files can be printed out for any defined area at any scale, and can provide single or multiple geological themes or layers specified by the customer. Thus it is possible, for example, to have a map showing made ground and the drift-free outcrop of a bedrock sandstone aquifer to determine pollution risk, or a map displaying only areas of outcropping coal seams indicating where historical shallow mining may present a hazard. There is also the good news that the printed output of all maps, even the 1:10 000 scale ones, will be in colour once more, making them more accessible to and readable by a wide spectrum of users.

The final advantage of the trend towards increasing digitization will be that levels of information from generalized to detailed will be readily available and accessible through PC discs, CD Roms and computer networks. This means that behind any map will be digital datasets of geological and other factors including borehole logs, landslip records and geochemistry which can be retrieved on demand simply and, we hope, cheaply by those seeking information about an area, from the site specific to the general.

Conclusion

Geological maps can provide many clues about urban settlements, from the reason for their establishment in the first place to the secret of their continuity and growth. Every city has a story to tell that is probably related to the geology. The maps are the key, and although there is a progressive move to customize maps through digital technologies for a multitude of uses, the use of computers will serve to enhance the scope of information that can be gleaned from them. The basic paper Solid and Drift geological maps may remain the most popular products, but behind them will lie a readily accessible reserve of supplementary information, held in digital databases. For the adventurous there will also be the scope to move into and use the third dimension on a computer screen and no longer rely on the imaginative leap that the present 2D map requires.

Acknowledgements

I am grateful to colleagues at the British Geological Survey for helpful discussions and comments over a long period on the matters addressed in this review chapter. The chapter is published with the permission of the Director, British Geological Survey (NERC).

References

Charsley, T. J., Rathbone, P. A. & Lowe, D. J. 1990. *Nottingham: A geological background for planning and development.* British Geological Survey, Technical Report, **WA/90/1**.

McAdam, A. D. 1994. *Edinburgh: A landscape fashioned by geology.* Scottish National Heritage, Perth.

Prosser, C. D. & Larwood, J. G. 1994. Urban site conservation - an area to build on? *In*: O'Halloran, D., Green, C., Harley, M., Stanley, M. & Knill, J. (eds) *Geological and landscape conservation.* Geological Society, London, 347-352.

Wilson, H. E. 1985. *Down to earth: one hundred and fifty years of the British Geological Survey*. Scottish Academic Press, Edinburgh.

3 The nature of the urban geological resource: an overview

Jonathan G. Larwood & Colin D. Prosser

Summary

- There is a strong link between human cultural development and the use of the Earth's geological resources.
- Geological resources have played a central role in urbanization and are part of the very fabric of our urban areas.
- The geological resource of urban areas is diverse, and reflects both the remnants of the pre-urban landscape, as well as the very fabric of the built environment

The effect of topography and the presence of the raw minerals of industrialization in determining the location of our towns and cities are clear. Equally, building stone is a reflection of local geology. Underlying geology is revealed as natural outcrops, from river cliff to sea cliff, and as man-made outcrops from quarries to road cuttings (the result of industrialization and urban development). This diversity makes urban areas some of the richest, most varied and accessible geological resources.

Urban areas provide a unique resource through which the links between human development and geology can be understood by the wider public. Ultimately, however, urban development can pose a threat to this geological resource. Conservation is needed and is vital to ensure that the diversity and scope of this urban resource are recognized and used. An increased understanding of urban geology and the need for geological conservation in our towns and cities, will form an important foundation for future geological conservation throughout the country.

Urban geology, in particular the origin and quality of building stone, has fascinated geologists since the last century. Charles Kingsley presented a series of lectures to the Chester Natural History Society in 1871 dealing with topics such as 'The Stones in the Wall' and 'The Slates on the Roof' clearly addressing the presence of geology in towns and cities. By the early part of this century Elsden and Howe (1923) had documented the variety of building stone found in London and today there are many guides and trails to the building stones of our towns and cities. The full range of the resource available, however, has yet to be fully recognized and the conservation of the urban geological resource is still in its infancy (Prosser & Larwood 1994). In our towns and cities, geology is all around us; it is almost a matter of being unable to see the wood for the trees.

Historically, our link with geology is inescapable. From primitive society onwards, the value of the geological resource has been without doubt. Whether for aesthetic or more practical reasons, our intellectual and cultural development has moved hand-in-hand with our understanding and exploitation of the geological environment, this being driven by the basic human priorities to obtain food, clothing and shelter (Shackley 1977). For example, the use of stone or exploitation of metal ores in making tools or the utilization of stone building materials to replace cave and rock dwellings mark significant cultural developments linked with our growing understanding of the surrounding geological resource.

Towns and cities bring this link home to modern society; their location, development and construction are reliant on the surrounding and available geological resource (Charsley 1996). The wider understanding and use of this resource, from local government to local community, is essential for the future conservation of geology in urban areas. This paper identifies the scope and value of the urban geological resource, identifying the need for its conservation and discussing means by which this can be achieved.

The cultural link

Every culture has demonstrated a value for the Earth's mineral resources. Such resources are clearly tied to our intellectual development from the earliest use of stone as a tool and weapon, through to the industrialization of the nineteenth century and today's global exploitation of mineral resources. Equally, fascination for the curious and beautiful has placed an aesthetic, and hence a monetary, value on our surrounding rocks, fossils and minerals.

Aesthetic value of geology

Rarity, beauty and distinctiveness have attached an aesthetic value to geological materials. Rare stones were valued and turned into precious jewels while precious metals have been melted since the Stone Age when the first gold jewellery was made. Such has been our desire for precious metals and stones that discovery of the Inca wealth drove the sixteenth century colonization of Central and South America by western civilization. Sporadic 'gold rushes' have occurred throughout history, the biggest, that of Yukon in 1896. Likewise, the discovery of South African diamonds at Kimberly in 1867 produced an unprecedented 'diamond fever' as people abandoned jobs and homes to dig for diamonds on the banks of the Vaal and Orange Rivers (Kourimsky 1977).

Similarly, fossils have adorned jewellery, been used for barter and become linked with folklore. The ammonites of the North Yorkshire coast, such as *Dactylioceras* and *Hildoceras*, have, since medieval times, been thought of as snakes, turned to stone by the Whitby abbess St Hilda and losing their heads as a result of St Cuthbert's curse (Shackley 1977). Many of these 'snakestones' were subsequently embellished by local dealers with carved snake heads.

Practical applications of mineral resources

The practical use of rocks and minerals has intrinsic links with our intellectual development. The use of rock for tools and weapons led to some of the earliest economic exploitation of mineral resources. An excellent example is the flint mines at Grimes Graves in Norfolk, now conserved as a geological and archaeological site of importance. The different techniques developed in flint manufacture, widespread across Europe, are today used as a basis for defining Palaeolithic cultures. At the same time, other rarer and distinctive minerals were being encountered and the first metals were being melted. Copper and then tin were discovered, and the chance mixing of the two, formed the more useful alloy bronze, leading to the dawn of a new era in human history, the Bronze Age.

Overcoming the problems of high temperature melting, some 3000 years ago, brought with it the Iron Age, initiating an unprecedented advance of human activity and development in agriculture and crafts. By the Middle Ages, iron was the most widely used metal, its production becoming one of the key manufacturing industries of the later Industrial Revolution. Witness the iron and steel industries exploiting iron resources such as the Cleveland, Frodingham or Northampton Ironstones.

Today it is impossible to imagine an existence that does not in some way use our surrounding geological resource. Our link with geology is inescapable and has provided a key driving force, from the earliest cultures, in our intellectual and cultural development.

Urbanization: the link with geology brought home

Topography

Topography has always driven settlement pattern. The need for a defendable vantage point led to the construction of hill-top forts and today many towns and cities have grown in the protective shadow of similarly placed castles. Where high ground proved inhospitable, towns and cities built up adjacent to fertile flood plains or in the shelter of upland valleys.

Durham City provides a classic example of the use of protective topography. In seeking sanctuary from marauding Vikings, monks from Lindesfarne took advantage of the natural protection provided by an incised meander in the River Wear. Not only do the cathedral and castle sit on a natural vantage point created by the regional uplift of Carboniferous sandstone, but the River Wear, as it has cut down through this sandstone, forms a natural moat.

Mineral resource

The link between urbanization and economic mineral resources is unequivocal and, in many cases, has led to the success or failure of continued urban expansion (Prosser & Larwood 1994; Charsley 1996). Early settlements developed around the flint mines of Grimes Graves and similarly in association with the copper and tin ores of Cornwall. Mineral resources such as iron, lead and tin were the primary economic incentive for the Roman invasion of Britain (Blagg 1990).

This link, however, is perhaps most tangible when considering the industrial development of the eighteenth and nineteenth centuries (Briggs 1979). By the end of the eighteenth century, coal mining was established in northeast England, initially to supply the demands of cities such as London. Many new industrial techniques were established using the available natural resource of the northeastern coal fields. The iron and steel industries flourished, and as a by-product of the coal industry, so did the glass trade using the raw materials and facilities of the coal industry for glass manufacture. This vast resource also stimulated the development of the steam engine for pumping mines as well as the invention of the earliest locomotives, bringing a new era in transport.

Industrial development linked with urbanization is mirrored in all areas where there was a naturally occurring and plentiful supply of coal and iron. As early as 1617 the claim was made that 'more people were living within a radius of ten miles of Dudley and more money returned in a year than in four Midland farming counties' (Briggs 1979); Dudley was already the heart of what was to become known as the 'Black Country'. Here the naturally occurring coal and iron of the South Staffordshire Coalfield came together to form another focus for industrial development and associated urbanization.

In Dudley itself, Silurian limestones were quarried to provide a flux for the growing iron industry. The quarries that were created are still visible today as vast caverns underneath Dudley Castle and open caverns in the Wren's Nest National Nature Reserve (Fig. 3.1). This area was subsequently made famous for its geology by Murchison (1859) who published his great work *Siluria*, in part based on the many observations made in quarries in and around Dudley.

The limestones of Wren's Nest have also long been renowned for their abundant Silurian reef fossils, perhaps most notably the trilobite *Calymene*. Pristine preservation placed a high value on fossils such as *Calymene*, which became affectionately known as the 'Dudley Bug'. So important has been the link between geology and industrial development in the Black Country that, for many years, Dudley Metropolitan Borough Council carried an image of the Dudley Bug on its coat of arms.

Fig. 3.1. Wren's Nest National Nature Reserve. Silurian reef knoll overlooking Dudley.

The built environment

The very fabric of our towns and cities is moulded by the underlying and surrounding geology. The Romans came to Britain in search of economically valuable minerals but also brought the wider use of stone as a building material, changing forever the built landscape of Britain (Blagg 1990). In fact, the Romans successfully sought out and exploited many of the stones which even today have subsequently been found useful in building (Dove 1996).

Buildings in Britain, until the end of the eighteenth century, strongly reflect the locally available building stone, difficulty of transport being a strong controlling factor in the wider distribution of building materials (Dove 1996). A traverse across England from southeast to northeast still reveals a variety of building stone which mirrors the changing lithologies of the underlying geology (Clifton-Taylor 1987). From the eighteenth century onwards building stones were moved further afield as transport barriers were overcome and towns and cities began to show a myriad of building stones of regional, national and even global origin (Dove 1996).

Peterborough City shows a building stone development typical of many areas. Early construction used locally available building stone, notably Lincolnshire Limestone. Most splendid of Peterborough's early surviving buildings is the cathedral (Fig. 3.2). The west front was constructed in the thirteenth century from the local Barnack Stone, the disused stone quarries of which now form the Hills and Hollows National Nature Reserve (Fig. 3.3) which is noted for its limestone flora and associated fauna. The affluence associated with the cathedral led to the import of other local decorative stones as well as those from further afield. Shrines were created in chalk from southeast of Cambridge and monuments carved from marbles including locally available Alwalton marble, and more distantly Frosterly Marble from Weardale in County Durham. The nineteenth and twentieth centuries saw an increased mixing of more exotic building stone in Peterborough. Shop and bank facias include numerous granites and the main Queensgate shopping centre is dominated by polished Jurassic limestones from Italy and Germany and orbicular granite from Finland.

Fig. 3.2. The west front of Peterborough Cathedral constructed from Barnack Stone (Photograph: R. Cottle).

Fig. 3.3. Disused quarries at Barnack, the source of Peterborough Cathedral's Barnack Stone. Now the Barnack Hills and Hollows National Nature Reserve (Photograph: J. G. Larwood).

Since the nineteenth century Peterborough has also had a strong association with the brick-making industry. Advances in the brick-making process from 1880 onwards (Clifton-Taylor 1987; Hillier 1981) and the discovery of the vast Oxford Clay resource - 'clay that [literally] burns' (Hillier 1981) - led to a commercial exploitation still important today.

The influence that geology has had, and continues to have, on urbanization is unparalleled. Whether through the exploitation of topography, mineral wealth or building stone, the link between geology and the development of our towns and cities is inescapable. For this reason our urban areas provide perhaps the best opportunity to demonstrate the value of geology, and in consequence, the need for geological conservation in modern society.

Today's resource and the need for conservation

The link between urbanization and geology has been outlined earlier, but in what form is the geological resource of our towns and cities expressed today?

- Natural exposures - here the topographical influence is often apparent. Natural exposures can include river sections, natural crags and outcrops and even coastal cliffs which so often reflect the original decision to site a town on a vantage point or beside a navigable river.
- Man-made exposures - these are very much the result of the urbanization process. They can include active and disused quarries and mines and also cuttings and tunnels created by expanding road, rail and canal networks.

Uniquely, however, towns and cities offer us the built environment itself. Building stones, though not in the traditional sense classed as exposures, add a unique geological element to our towns and cities, reflecting local geology as well as a wider geological story which may otherwise be inaccessible.

Despite the legislative protection offered through designation as National Nature Reserves, Sites of Special Scientific Interest, and Local Nature Reserves, and in recent years recognition as Regionally Important Geological/Geomorphological Sites (Harley 1992), urban geological outcrops are still, by their very nature and location, under continual and increasing threat from development.

Undeveloped land is at a premium for living space as the knock-on effects of urbanization are many; though the advantages for the creation of new exposure are obvious, this same exposure can with equal speed be destroyed for ever. Growing population leads to an ever increasing demand for waste disposal and landfill, the potential effects of which can even be seen as far back as sixteenth century in Norwich. Here chalk quarries, once dug for building material, were already filling up as a convenient repository for the city's waste (Ayers 1990). Although the highest pressure on our geological resource is undoubtedly in urban areas, the very diversity of urban geology offers our best opportunity for the innovative conservation and promotion of our earth heritage. What steps can we take towards seeing this potential?

1. Recognition of the scope of the resource. As outlined in this paper our towns and cities provide a wide range of exposures, both natural and artificial. Urban areas, however, offer the opportunity to look beyond the classic rock outcrop to the actual fabric of the built environment from which much can be learnt about local, national and even global geology.
2. Use and documentation of this resource. Urban areas provide us with an accessible geological resource that can be used by schools and university students alike and equally enjoyed by the member of the general public as well as the amateur and professional geologist. It is important to document this resource fully. Many towns and cities have established record centres (Reid 1994; Pounder 1996) documenting existing statutory and non-statutory sites. Reid (1996) outlines a system for taking this documentation one stage further in the hope of involving all developers affecting the urban geological environment. Town geological trail guides can also provide an accessible insight to this resource, not only documenting outcrops and setting them in a wider geological context, but providing a guide to building stone (e.g. Robinson 1984, 1985; Jenkins 1988; Dove 1994). In recent years museum innovation has striven to popularize geology as a subject, presenting geological information in an accessible and exciting fashion as well as drawing the link between local history and local geology (Reid 1994; Knell 1996).
3. Value the resource. If documented and used, then the value placed on our urban geological resource can only increase. Geological conservation in the Black Country illustrates well how this can be achieved. The conservation value of many geological sites in this area is recognized by their formal designation as both statutory and non-statutory sites (Box & Cutler 1988; Cutler 1996; Reid 1996). This status is recognized within the Black Country Strategy (Black Country MBCs & English Nature 1994) which offers a commitment to the conservation and enhancement of this resource. Alongside this, the Black Country Geological Society has done much to promote the local value of geology to the area, and together with Dudley Museum and Art Gallery has provided the link between local geology and the industrial heritage of the region, the link of greatest value to the local community.

Perhaps most important here is the establishment of a formal strategy recognizing the need for geological conservation and setting out means by which this can be achieved. Durham County Council (1994) has produced the first strategy devoted to geological conservation, *Geological Conservation Strategy*. This outlines the need to document and periodically review the resource, but most importantly promotes 'the creation of new geological sites at scientifically important horizons' and 'suitable geological and geomorphological sites for teaching purposes'.

Conclusion

The link between human cultural development and geology, whether aesthetic or practical, is undeniable . This link to society is the key to raising awareness of geology and the need for its conservation. The rich and varied geological resource in urban areas presents perhaps our best opportunity for achieving this goal through involvement at all levels.

References

Ayers, B. S. 1990. Building a fine city: the provision of flint, mortar and freestone in medieval Norwich. *In*: Parsons, D. (ed.) *Stone quarrying and building in England, AD 43-1525*. Phillimore, London, 217-228.

Black Country MBCs & English Nature. 1994. *Black Country Nature Conservation Strategy*. English Nature, Sandwell MBC, Dudley MBC, Walsall MBC and Wolverhampton MBC.

Blagg, T. F. C. 1990. Building stone in Roman Britain. *In*: Parsons, D. (ed.) *Stone quarrying and building in England, AD 43-1525*. Phillimore, London, 33-50.

Box, J. & Cutler, A. 1988. Geological conservation in the West Midlands. *Earth Science Conservation,* **25**, 29-35.

Briggs, A. S. A. 1979. *Iron Bridge to Crystal Palace: impact and images of the industrial revolution.* Thames & Hudson Ltd, London.

Charsley, T. J. 1996. Urban geology: mapping it out. *This volume.*

Clifton-Taylor, A. 1987. *The Pattern of English Building* (4th edition). Faber & Faber, London.

Cutler, A. 1996. The role of the regional geological society in urban geological conservation. *This volume.*

Dove, J. 1994. *Exeter in stone*, Thematic Trails, Oxford.
---- 1996 Exeter and Norwich: their urban geology compared during medieval, Victorian and Edwardian periods. *This volume*.
Durham County Council. 1994. *County Durham Geological Conservation Strategy*.
Elsden, J. V. & Howe, J. A. 1923. *The Stones of London*. Colliery Guardian Co., London,
Harley, M. 1992. RIGS update. *Earth Science Conservation*, **30**, 22-23.
Hillier, R. 1981. *Clay that burns: a history of the Fletton brick industry*. London Brick Company, Peterborough.
Jenkins, P. 1988. *Geology and buildings of Oxford*. Thematic Trails, Oxford.
Knell, S. J. 1995. Museums: a timeless urban resource for the geologist? *This volume*.
Kourimsky, J. 1977. Man and Stone. *In*: Lucas, R. (ed.) *The illustrated Encyclopaedia of minerals and rocks*. Neographia Martin, Czechoslovakia, 9-24.
Murchison, R. I. 1859. *Siluria* (3rd Edition). John Murray, London.
Pounder, E. 1995 Geomorphological conservation: opportunities afforded in Greater Bristol. *This volume*.
Prosser, C. D. & Larwood, J .G. 1994. Urban site conservation - an area to build on? *In*: O'Halloran, D., Green, C., Harley, M., Stanley, M. & Knill, J. (eds) *Geological and Landscape Conservation*. Geological Society, London, 347-352.
Reid, C. 1994. Conservation, communication and the GIS: an urban case study. *In*: O'Halloran, D., Green, C., Harley, M., Stanley, M. & Knill, J. (eds) *Geological and Landscape Conservation*. Geological Society, London, 365-371.
---- 1995 A code of practice for geology and development in the urban environment: a new Local Authority initiative. *This volume*.
Robinson, E. 1984. *London - illustrated geological walks. Book 1*. Scottish Academic Press, Edinburgh.
---- 1985. *London - illustrated geological walks. Book 2*. Scottish Academic Press, Edinburgh.
Shackley, M. 1977. *Rocks and Man*. George Allen & Unwin Ltd, London.

4 The changing nature of the urban earth heritage site resource

Susan M. Ingham, Peter Doyle & Matthew R. Bennett

Summary

- A survey of the *Proceedings of the Geologists' Association,* published over the last 100 years, demonstrates that the nature of urban sites exposing geology within south London has changed dramatically.
- Urban growth, particularly during the last 50 years, has led to marked deterioration of natural exposures, with periodic fluctuations in artificial exposures.
- This marked deterioration emphasizes the importance of artificial geological resources in increasing public awareness of geology in the urban environment.

The changing nature of the urban geological resource is difficult to assess. Urban growth, particularly in large conurbations such as London, occurred rapidly with the development of an efficient transport system between 1900 and 1939 (Jackson 1991). Large areas of once rural countryside were swallowed up in the creation of outer suburbs. The nature of urban development is such that it both creates and destroys geological sites. For example, development reveals new geological exposures during the construction of cuttings and in the extraction of minerals such as aggregates. Some of these exposures are quickly infilled or lost, while others remain and provide urban geologists with a valuable resource. Development may also destroy geological sites, particularly as the pressure for land increases within urban areas. This paper describes a simple exercise which attempts to document the changing nature of the site resource within south London, over a period of a hundred years from 1880 to the present day.

Methodology

South London was chosen as the study area because it is constrained by the Thames in the north and the boundaries of the London boroughs to the south, and because it contains a record of most of the periods of urban growth up to the present day, together with a varied site resource.

The *Proceedings of the Geologists' Association* (*PGA*), first published in 1859, has a long established history of recording the activity of local geological groups particularly in and around London (Middlemiss 1989; Robinson 1989). It frequently contains reports on field meetings and site based research reports (Green 1989). As a journal it provides an excellent record of the activity of both amateur and professional geologists alike. It can therefore be used to record the history of individual geological sites.

A systematic survey of each volume of the *PGA* since 1880 was undertaken in order to build up a site database for south London. This database was derived from two types of information: (1) research papers which referred to temporary or permanent sites present at the time of publication; and (2) field meeting reports of excursions to temporary or permanent sites present at the time. These sites were classified into permanent or temporary exposures. Temporary sites were defined as any temporary excavation during development work which was of limited duration and extent. Permanent sites were defined as any permanent natural geological features such as landforms and quarries, pits and cuttings at which the exposure is known not to have been of limited duration or extent. The number of sites mentioned within the pages of the journal shows a marked change since 1880. The advantages of using the *PGA* as a primary data source are: (1) the bias of the Geologists' Association membership to the geology of South East England (Middlemiss 1989); (2) the long run of its journal and the consistency of standard of its published papers and their style/format; (3) the range of geological and geomorphological topics covered; and (4) the range of interest groups catered for by the journal.

The disadvantages and limitations of the *PGA* dataset are: (1) that papers will reflect changing fashions and interest within the subject leading to research papers and field reports; and (2) the absence of an independent test or other data set against which to test its reliability.

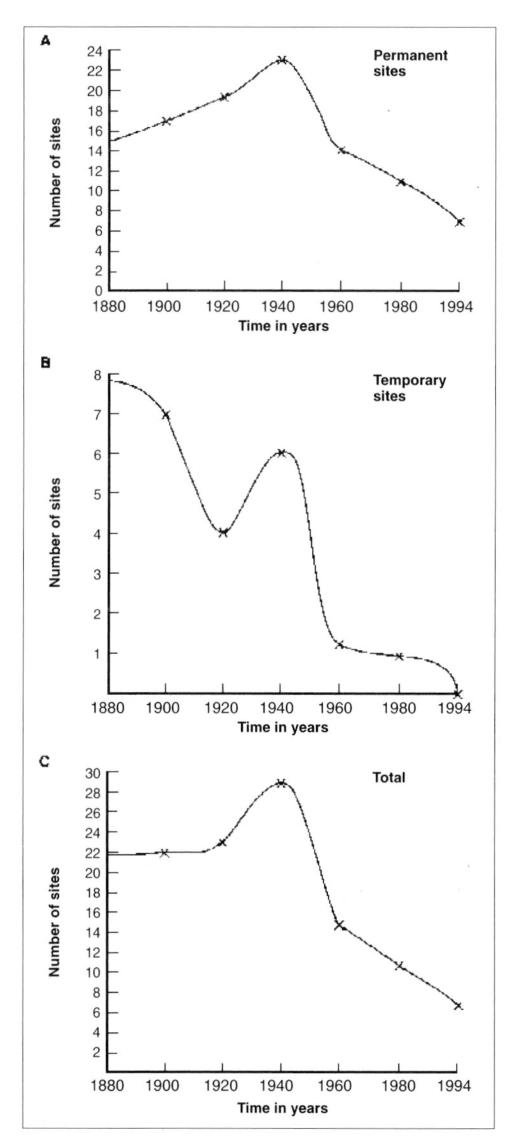

Fig. 4.1. A series of three graphs showing the changing nature of the site resource within south London. See text for details of their construction.

Despite these limitations, the survey should reflect the changing nature of the site resource, and as such it can be used a proxy record for the changing number of sites within London. The pages of the *PGA* were scoured and the titles of the prominent papers recorded for decade blocks commencing for the decade between 1880 and 1890 and continuing to the present day. The nature of the site temporary or permanent was recorded. These data were compiled into three simple graphs (Fig. 4.1).

Results

Figure 4.1a shows a graph of the number of permanently exposed sites mentioned within the pages of the *PGA*. The graph suggests that the resource of permanently exposed sites reached a maximum in 1940 and has declined thereafter. Figure 4.1b shows the record for temporary exposures, which shows a gradual reduction in number from 1880 to 1994. This is probably reflective not of the number of temporary exposures but of the frequency with which they are documented within the pages of the *PGA*. This may be a function of the fact that when the subject was young, almost every temporary exposure added to our level of knowledge, but that today a temporary exposure has to contain something remarkable or special before it is written up in a paper. This graph illustrates an important point: the rise in temporary exposures during the 1920s and 1930s, which corresponds with the one of the most dramatic periods of urban expansion within south London (Jackson 1991). Figure 4.1c combines both the temporary and permanent exposure sites within London and illustrates the apparent decline within the resource base of natural geological sites within this area. It is a decline which reflects the gradual loss of quarries to development and to general site degradation which has taken place within south London.

Discussion

Although crude, the *PGA* dataset clearly demonstrates the deterioration of the site-based resource within south London. Despite the fact that many once rural areas with quarries and other geological exposures were swallowed up during the expansion of London in the 1920s and 1930s, the pressure for development led to the demise of permanent sites, although promoting some temporary exposures. This exercise demonstrates two things: that within urban areas, naturally occurring geology diminishes with time; and that with degradation of the natural resource the need for artificial geological resources - building stones, parks and museums - within urban areas increases. This is particularly important given the inclusion of geology in the National Curriculum (Hawley 1996). We can attempt to conserve the few natural exposure sites which exist in London through active conservation, but this exercise illustrates the need to consider alternative resources such as those provided by building stones, museums and parks in the promotion of geology.

References

Green, C. P. 1989. Excursions in the past: a review of the Field Meeting Reports in the first one hundred volumes of the *Proceedings*. *Proceedings of the Geologists' Association*, **100**, 31-54.

Hawley, D. 1996. Urban geology and the National Curriculum. *This volume*.

Jackson, A. A. 1991. *Semi-detached London*. Wild Swan, Oxford.

Middlemiss, F. A. 1989. One hundred volumes of geology: a personal review of the *Proceedings* since 1859. *Proceedings of the Geologists' Association*, **100**, 55-72.

Robinson, E. 1989. The origin and early years of the *Proceedings*: context and content. *Proceedings of the Geologists' Association*, **100**, 5-16.

Part Two

The nature of the urban geological resource

This part of the volume documents in four sections the nature of the urban geological resource. The first, ***Building stones as an urban geological resource***, demonstrates the value and scope of building stones in the urban environment. The second, ***Parks and open spaces as an urban geological resource***, illustrates that parks are important in promoting an awareness of geology and geomorphology in a semi-natural state. The third, ***Museums as an urban geological resource,*** goes some way in documenting the rise and importance of museums in promoting urban geology. The fourth, ***Urban geology and civil engineering***, gives two demonstrations of the role of civil engineering in urban geology, both through temporary exposures, and in the use of engineering principles to help conserve unique components of urban geology.

5 'The paths of glory . . .'

Eric Robinson

Summary

- The opportunities for fruitful urban geology are numerous, but they need to be sought out and pursued with enthusiasm.
- A recent survey of Hoxton Square in London illustrates what can be achieved if local geologists become involved with local schools.
- Cemeteries have great potential as urban geological resources, and may be an area where volunteer conservationists have an important role to play.

In their work with Common Ground, Sue Clifford and Angela King make a point of stressing what they call local distinctiveness within the English landscape; that is, all the aspects which combine to make up the total character of villages and even whole regions. Best seen in their work on the 'English Apple', it is a concept which can extend to geology. There is a sense in which the complex geological map of England is reflected in buildings and their building stones, in gravestones and paving, in roofing materials and in walls. For urban areas, the story is at the same time both simpler and more complex. A certain familiarity is associated with the Victorian and Edwardian eras when the same rock types were used time and time again. For recent years, however, it can be a different matter; all hell broke loose in the post-war years as buildings became clad in Namibian gneiss or Brazilian yellow granite. The important fact remains, however, that *geology is all around us* if we care to look. This contribution aims to point out the opportunities which are there to be taken or ignored as we choose.

The improved environment

Recently, I was contacted by the Environment Services Department of the London Borough of Hackney to identify the stones which surround the green space of Hoxton Square. These stones once provided the base for substantial railings. Laid out in the eighteenth century, only two of the original buildings survive, but the Square still has at its core a welcome extent of grass lawn with some mature London plane trees. Some years ago, the railings cut down in war-time for scrap metal, were replaced by post-war substitutes which are too flimsy and are due once more to be replaced. But what about the stone bases? Were they in good condition? What kind of stone were they? These are the simple questions, questions which any geologist could respond to, but the key, however, is to think how the answers can be expanded into suggestions which could benefit the community.

The stone bases to the old railings in the square prove to be 1-1.5 m lengths of Cornish Granite, little different from city kerbstones, but each length is perforated by ten or twelve square holes which took the railings. Apart from some slightly drunken alignments, little needs to be done to renew the original stone bases. It could rest at that, but nearby in the square other geological details become apparent (Fig. 5.1).

First, the actual kerbstones. All around Hoxton Square the kerbstones of the inner-site are almost entirely of bluestones. Speckled blue-grey in colour, with occasional straggling white veins crossing their surface, these are good examples of Guernsey Diorites, a kind of stone which was very popular with London parishes in the nineteenth century when they began to provide pavements for pedestrians safe from the horse-drawn carts which crowded the roads. Ready-dressed in the quarries of the island, these stones came through the Port of London in vast quantities. To the geologist, the interest in these Ordovician igneous rocks lies in the interpenetrating intrusion of darker and lighter varieties of diorite.

The outer pavement kerbstones, particularly those on the north side of the square outside St Monica's Church, seem in contrast, to be more often of Cornish Granite and often show the floods of white feldspar laths which typify such granites. Other kerbstones have a dull red tone typical of the Mountsorrel Granite from Charnwood Forest, Leicestershire. These stones would probably have been brought south to Hoxton by the Grand Union Canal which runs close to the present site on its way to the

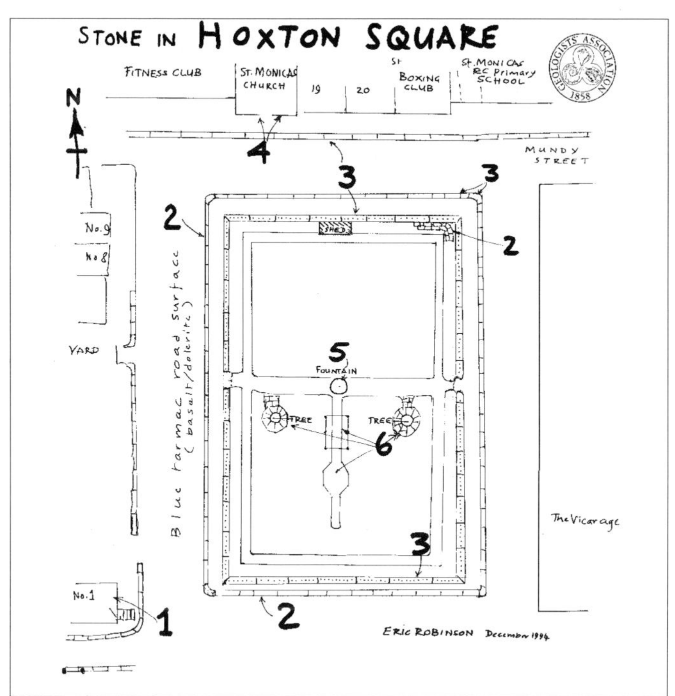

Fig. 5.1. Manuscript plan of Hoxton Square showing the wealth of its geological resource. (1) Blue plaque to James Parkinson (1755-1824), a geologist who collected fossils from the London Clay. (2) Bluestone kerbs: stone from Guernsey in the Channel Islands. Diorite is the name of this stone. (3) Grey granite base slabs to the original railings around the square gardens. Probably from cornwall. Kerbs on the north side outside St Monica's. (4) Bath stone: yellow limestone with much broken shell forming the flakes which stand out and making the stone rough to touch. (5) Grey granite in the bowl to the fountain in the centre of the square. From Aberdeen. (6)Pavings of sandstone, probably from Yorkshire quarries. Notice the thin layers and marks on the surface which are fossil ripple-marks made by flowing water.

Thames at Shadwell. The same lithology crops up in the cobbles which pave the entrance to a yard on the west side of the square. Here, there are also several gneiss cobbles of Scandinavian origin.

The pavements themselves are a bit disappointing - often simply tarmac surfaces - but what we might have hoped for, York Stone flagstones, are to be found inside the central garden where broken pieces surround the bases of the two great plane trees, and also make the path which runs through the pergola walk which forms the central axis to the southern end of the garden. In the right kind of incident light, you can persuade yourself that there are faint ripple-marks running across the bedding surfaces. Certainly, you can see the thin laminations which make up the flagstones and give the rock its splitting quality which makes it an excellent paving slab.

Still in the central garden, the next geological detail is the truncated base of a drinking fountain put there in 1902 by the Passmore Edwards family as one of their many good works in the East End of London. From our point of view, it adds a sombre grey Aberdeen Granite, probably Rubislaw Granite from the City of Aberdeen, to the range of granites which are to be found in the square. Because of vandalism, a broken surface to the bowl allows us to see silver flakes of muscovite mica as well as pitchy black wisps of biotite. If the fountain is in fact restored to working order as is proposed as part of the improvements to the square, it would be worthwhile having the granite cleaned of the dark coating of grime, if only to reveal a little more of the typical Caledonian granite to contrast with that of southwest England.

Looking up from the ground, the buildings are an odd mixture of styles. The only surviving town house with Georgian proportions is to be found on the east side of the square and has been recently restored as office premises. Of the later Victorian modifications to the square, most relate to St Monica's Church and the diverse buildings which were associated with its Catholic mission work in Hoxton and south Hackney: a school (now a fitness club), chaplaincy and mission hall. All, including the church with its spiky Gothic frontage, are fashioned in yellow London Brick with thin bands of contrasting red and black. Associated with the brickwork there are string-courses and copings of pale yellow Bath Stone. Suitably weathered after 100 years of exposure, the surface of this limestone is rough to the touch thanks to countless shell fragments which stand proud of the corroded surface. With the opportunity to introduce fossils, sedimentary bedding and the magic name Jurassic this small outcrop is of particular importance.

There remains in the southwest corner of the square an undistinguished building of glazed brick, which stands on what was the site of the London home of James Parkinson (1755-1824). In his day, he was one of the greatest promoters of our science. His great fascination was the London Clay and the fauna and flora which came from it in the course of the many civil engineering projects which were going on in London at this time. This interest was well known to the local people to the extent that he was the subject of an early cartoon making gentle fun of a geologist, suggesting that his views were at least being discussed, if not totally understood. Today, the same James Parkinson would be better remembered for his recognition of the tragic nerve disease which carries his name, but the blue plaque records him jointly as 'physician and geologist'.

The educational potential within Hoxton Square is considerable (Fig. 5.1). Knit together these observations and they could constitute a geological trail which could explore most of the basic principles of geology which are required in the National Curriculum. In the top northeast corner of the square stands St Monica's Primary School. Proud of a recent environment awareness award, its pupils have to become aware of crustal materials and the processes of weathering as part of the National Curriculum. I can think of no better use of a geologist's time than to write up that geological trail or to devise the necessary projects for the different age groups. Figure 5.1 represents a simple plan of the geological resource within the square and illustrates what can be done to provide a teaching resource in an urban area.

To the pleasure gardens south of the river

Historically, the Battersea riverside has always proved to be a magnet for society. A Pleasure Garden in Stuart and Georgian London, it became so again in 1951 during the Festival of Britain year. Today, this neglected park deserves the attention of any urban geologist intent upon spreading the influence of geology or once again, providing the means of study for local schools.

The originally marshy ground of the 200 acre site was in part raised by the dumping of silts excavated when the Surrey Docks were created in Bermondsey. Our geological opportunities in the park arise from another consequence of the trade into the Port of London. Sailing vessels of the time were always dependent upon the stability provided by stone ballast to be able to sail effectively. On entering port, they invariably discharged their ballast, tons of beach boulders or quarry waste. In the Thames these piles of ballast accumulated in the Greenwich reach. It was material which went into road building, but in this case, was used to build the walls which protected the nursery gardens of the Park in its northwest corner close to the Albert Bridge. As we see them today, they are a splendid mixture of colours as limestones from Portland and Purbeck sit alongside Kentish Ragstone, Cornish Granites and the range of yellow-brown sandstones brought south from the Yorkshire coalfield.

Colour, including stone reddened by fire, is a good guide to true difference which can be recognized by pupils from any primary school if we are not too demanding of proper identifications. Teachers can come with pupils of a higher level of attainment, all the way up to A-level and beyond. They would not be deceived by a modest trick. Short on volcanic rocks, some perfectly good and vesicular slags from Thames-side foundries fit the need, but would be discriminated by more accomplished pupils. You still must expect some measure of deception in this most famous of pleasure gardens. This example shows the surprising variability of the urban geological resource. The educational value of this resource is fully explored elsewhere in this volume (Robinson 1996).

Those paths of glory

If you read the poem *Elegy written in a country churchyard* by Thomas Gray, from which my title is taken, you will discover that my text comes at the end of a stanza which reads in full:

> 'The boasts of heraldry, the pomp of power,
> And all that beauty, all that wealth e'er gave,
> Awaits alike th' inevitable hour,
> The paths of glory lead but to the grave'

Inevitably, I could not speak of urban geology without ending with a comment upon gravestones and the work which awaits us in urban churchyards and cemeteries. It is here that undoubtedly our greatest openings are to be found. My own interest began way back in the early 1960s and was for many years a curiosity over the bizarre and eccentric memorials which people had raised to commemorate the dead. In the last ten years, however, opportunities have given a purpose to what my friends have long thought of as eccentricity.

First, there has been the cause pursued by the conservationist/historians led by The Victorian Society. Through the work of this society cemeteries throughout the country have been adopted by local groups. 'The Friends of . . .' has been a movement which has grown apace especially in inner city areas. Sustained by genealogists, art historians and those interested in purely local history, all have found need to call for the assistance of a geologist. Kensal Green, Highgate, Abney Park, and most recently Tower Hamlets have claimed my time. Articles in the *Geologists' Association Circular*, *Natural History* and *Geology Today*, as well as workshops for the Earth Science Teachers Association (ESTA) and RockWATCH have done something to spread the challenge to others.

Latterly, it has been the requirements of teachers tackling the National Curriculum, which have given a new purpose and justification to the surveys and the trail blazing. What we can do on the streets and even in public parks, can be done with greater safety in cemeteries, making these places the best option for working with schools. What is more, the school itself can become involved through schemes of conservation in which the pupils undertake clearance and upkeep. This is a situation which is developing at present in the churchyard of Kings Norton in Birmingham. Across the road from St Nicholas churchyard is a junior school which last year was rated 'Environment School of the Year'. The Headmistress, Mrs Field, is keen to maintain that success through involvement in gravestone trails which the parish has commissioned from a landscape architect. If all goes well, the local geological expertise will come from the members of The Black Country Geological Society. This in fact is a necessary part of such activity. Called in as an expert, I do depend upon others who live locally to sustain any scheme which might emerge, pressing this role upon any of the 15 local groups of the Geologists' Association, or those, such as the Black Country Geological Society, who are affiliated to the Association. In this way, geology groups looking for a local cause with community significance need look no further.

Other schemes which should bear fruit in months to come are already well in hand at Kensal Green, where Kensington Borough, British Waterways and the Friends of Kensal Green Cemetery are discussing what might be done through a City Challenge Award. In Tower Hamlets, a generous funding of the Soanes Centre in the corner of the now-closed Victorian cemetery, provides a local base for a community effort which might usefully be adopted by student groups of London's many universities and for the benefit of schools in Tower Hamlets and Stepney. In nearby Hackney, work for local schools has been provided by the team which manages the 32 acres of Abney Park Cemetery. Joining these existing projects we now have the Borough of Southwark seeking to use the Victorian cemeteries of Camberwell and Nunhead as part of their environmental work for schools in south London.

Time well spent?

Believing as we do that truly *geology is all around us*, we nevertheless need to go out and claim a role in any opportunity in the urban environment, like those which I have mentioned above. It is hopeless to wait for opportunities to come to us. We have to seek out actively our chance to perform.

We need to dragoon our clubs and societies to participate. To change from being passive conservationists with good intentions but negligible performance to those of deeds and action.

There is a need to persevere. Many efforts in conservation fail, but that is good reason to be professional in pressing support of good schemes, remembering the support which can be called upon. English Nature will often consider contributing to the running costs of such schemes if approached in good time. Then there is the possibility of support from the Curry Fund of the Geologists' Association for details such as signboards, interpretative boards or the production of explanatory trail leaflets. Local businesses are also an important source of support. Banks and supermarket chains, building societies and Rotary Clubs should be given the opportunity to support local initiatives which might carry their names and logo.

The best return for all this effort, other than the health of your local geological society or group, is the enthusiastic response which comes from the teachers and their pupils who make use of the resource that such schemes provide. It is this, above all else, which makes it all worthwhile.

References

Robinson, E. 1996. A version of 'The Wall Game' in Battersea Park. *This volume.*

6 Heathen, xenoliths and enclaves: kerbstone petrology in Kentish Town, London

Roger Mason

Summary

- The kerbstones in of our city streets have great potential for petrological study.
- Granite kerbstones from Kentish Town, London, are used to illustrate the important distinction between enclaves and xenoliths. This distinction is often difficult to make in the field.
- This study illustrates the potential of kerbstones and other stone blocks in urban areas for the study of petrological problems which are often difficult to study in the field.

Kentish Town is a typical inner London suburb. It is situated in the London Borough of Camden, 1.5 miles northwest of Kings Cross Station, and approximates to the NW5 postal area, between the better known districts of Camden Town and Highgate. A busy road from central London runs along the High Street, and the district is crossed in different directions by railways, some on elevated viaducts, some in cuttings and some underground (Anon 1991). Kentish Town lies in the valley of the 'lost' river Fleet (Fig. 6.1), which rises on Hampstead Heath, and flows in underground pipes past Kings Cross to join the Thames by Blackfriars Station (Barton 1992; Trench & Hillman 1993).

The history of Kentish Town has been described by Tindall (1977) in a classic of local history writing. A few buildings in the High Street date from the eighteenth and early nineteenth centuries, while the surrounding fields were built over in mid-Victorian times. As a consequence, most buildings are Victorian.

Many of the original building stones survive from this era. Close to shop fronts there are often original York Stone paving slabs, although against the carriageway they have usually been broken and replaced by concrete slabs or bricks. The Victorian road-surfaces of 'granite' setts (which are in fact various igneous rock types) have usually been removed or covered with tar macadam during road improvements, although they

Fig. 6.1. View northeast under an arch of the North London Railway on Prince of Wales Road NW5 (after a sketch described as a typical Kentish Town view by Tindall 1977). Several kerbstones illustrated are on this stretch of road. The brick building with a small spire to the left of the road is Kentish Town Swimming Baths, built in 1912. The eastern branch of the lost River Fleet passes beneath the road just beyond the baths, and the second car approaching on the right is on the crest of a rise between two branches of the river.

Fig. 6.2. The corner of Prince of Wales Road and Hadley Street. Victorian kerbstones of diorite (left and centre) and pink granite (right).

Fig. 6.3. The corner of Prince of Wales Road and Healey Street, closed by a traffic management scheme. The narrow kerbstones of white granite (in front of spray gun) and diorite (beyond the spray gun) are modern, as are the bluestone setts in the foreground, but Victorian York Stone paving slabs have been retained.

can be found in entrances and in one or two places where the road surface has worn through. Unlike the pavement slabs, most of the kerbstones have survived, although realignment of pavements and closing of roads for traffic management schemes in the last 15 years have introduced some new ones, which are easily distinguished from the Victorian ones because they are narrower (Figs 6.2 and 6.3).

The purpose of this paper is to demonstrate that the geology of the urban environment may not only represent a valuable educational resource, but also aid in scientific research. Kerbstones in particular are its focus. Traffic passing for more than a century has smoothed the kerbstones, whose surfaces were originally deliberately roughened by chisling (Fig. 6.5). Today's traffic keeps them smooth and relatively clean, and their petrography is well displayed for examination by the naked eye or with a hand lens. A spray bottle of water and a small stiff brush help to clean the loose dirt off, but anyone trying it should be warned that they will have to explain what they are doing to passers-by!

Kerbstone rock-types

Victorian and modern kerbstones are composed of three distinct rock types. These are pink granite, white granite and greenish diorite (Figs 6.2 and 6.3). The pink granite resembles Peterhead Granite from Aberdeenshire, the white granite similar to those commonly occurring in Devon and Cornwall, while the diorite appears to be the 'bluestone' often found as paving in London, and comes from the Channel Islands. These rock-types are described below.

Pink granite

This type of granite is only found as Victorian kerbstones. It contains crystals of pink orthoclase feldspar and white plagioclase feldspar up to 30 mm in length, with interstitial grey quartz and smaller crystals of biotite mica (Fig. 6.4). The feldspar crystals may display a characteristic euhedral lath shape and sometimes rapakivi texture, with white feldspar rimming pink.

Fig. 6.4. Pink granite, corner of Prince of Wales Road and Hadley Street. Some pink orthoclase feldspar crystals have rims of white plagioclase feldspar (rapakivi texture).

Fig. 6.5. White granite, Anglers Lane, with an unusually light coloured enclave barely distinguishable from the granite.

White granite

This is found in both Victorian and modern kerbstones. It contains only white feldspar crystals (Fig. 6.5), but there is both orthoclase and plagioclase. Some kerbstones show porphyritic texture, with orthoclase feldspar crystals up to 50 mm in length. Quartz is grey and interstitial, and when the weather is sunny, muscovite mica can be seen sparkling, as well as the dark biotite.

Diorite

This is found in both Victorian and modern kerbstones. It is more heterogeneous than the granites, with varying proportions of hornblende and plagioclase feldspar. Figure 6.6 shows diorite in a modern kerbstone with crystals of hornblende showing characteristic diamond-shaped cross-sections.

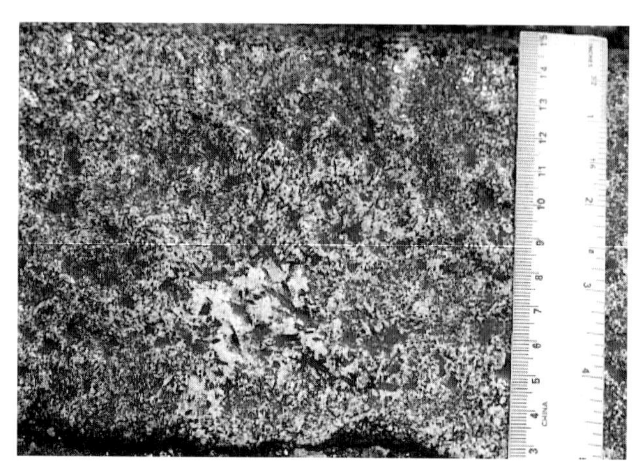

Fig. 6.6. Diorite, Prince of Wales Road by the swimming baths. Hornblende crystals display diagnostic prismatic and basal cross-sections in the pegmatite patch.

Enclosures in the kerbstones

The kerbstones of Kentish Town contain a variety of enclosures which are darker than the host granite or diorite. The terminology for describing these structures in English is considered by experts to be confused (Didier & Barbarin 1991; Pitcher 1993). In a comprehensive review Didier & Barbarin (1991, pp. 19–23) list the following names which have been used for such features: enclosure, knot, dark spot, bunch, nest of minerals, and most frequently, inclusion. Oddly enough, they omit a common quarrying term for dark enclosures in general, which is heathen. Pitcher (1993) agrees that the term inclusion should be reserved for enclosures within single crystals (for example fluid inclusions), and I shall follow his recommendation and avoid the word. Didier & Barbarin propose a consistent scheme of nomenclature using the general term enclave for all enclosures, as in French, but Pitcher points out that usage in English language publications (e.g. Middlemost 1985) does not fit in with this.

By contrast, an earlier generation of British petrologists, both transformist (i.e. those who considered granites to be created through transformation, such as Read (1957)) and magmatist (i.e. those who considered granites to be a primary product, such as Nockolds (1932), and Thomas & Smith (1932)), used xenolith for darker enclosures in general, preventing the word being used in a more precise sense, as in

Table 6.1. Two types of heathen (Modified from Didier & Barbarin 1991).

Term	Nature	Contacts	Shape	Other Features
Xenolith	Fragment of solid country rock	Sharp	Angular	Contact metamorphic minerals & texture
Enclave	Blob of coeval magma	Mostly sharp	Usually ovoid	Fine-grained igneous minerals & texture

Didier & Barbarin's scheme. I shall compromise between England and France by using the term heathen for the mafic enclosures in general, dividing them into xenoliths and enclaves. The latter term corresponds to the limited class of mafic microgranular enclaves defined and discussed by Didier & Barbarin (1991). I use their definition of xenolith (Table 6.1).

Xenoliths

Xenoliths only occur in diorite kerbstones and are rarer than enclaves. They show rectangular outlines (Figs 6.7 and 6.8) and a foliation fabric picked out by quartz veins preserved from the country rocks before incorporation into the granite. An even-grained hornfelsic texture can sometimes be recognized (Fig. 6.8). The xenolith in Fig. 6.8 has developed a thin rim of biotite mica, both at its contacts with the diorite and against the quartz veins. This is a characteristic feature marking the onset of partial melting of the block of country-rock, and melted xenolith material may be the cause of the lighter colour of the diorite against the right hand side of the xenolith in Fig. 6.8. There is a larger patch of lighter diorite around the xenolith shown in Fig. 6.7.

Enclaves

Enclaves occur in all three types of kerbstone; pink granite (Fig. 6.9), white granite (Fig. 6.10) and diorite (Fig. 6.11). Some are much darker than the surrounding host igneous rock (Fig. 6.9), while others show less contrast (Figs 6.5, 6.10 and 6.11). Most show the ovoid shape described by Didier & Barbarin (1991) but a few are angular (Fig. 6.11). All the

Fig. 6.7. Xenolith in diorite, corner of Prince of Wales Road and Grafton Road. Veins of quartz and feldspar pick out a pre-intrusion metamorphic foliation. The diorite is lighter coloured around the xenolith.

Fig. 6.8. Xenolith in diorite, corner of Patshull Road and Kentish Town High Street. A dark rim of biotite mica surrounds the xenolith and also fringes the quartz veins which pick out the pre-intrusion foliation. The material of the xenolith shows an even-grained hornfelsic texture.

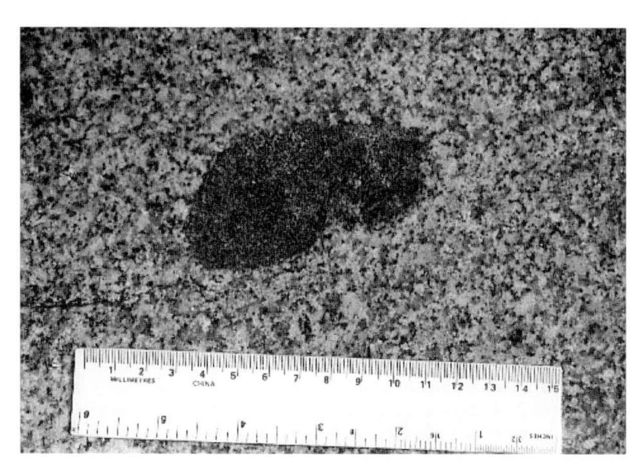

Fig. 6.9. Mafic enclave displaying igneous texture in pink granite, corner of Grafton Road and Prince of Wales Road.

enclaves display igneous textures, which are invariably finer-grained than those of the surrounding host rock. As Vernon (1984) explained, the features of enclaves make it virtually certain that they were molten magma of more basic composition than the surrounding host magma at the time of incorporation, and this can clearly be seen in examples in Kentish Town.

The ovoid shape of the enclaves shows that there was an excess surface energy at the contact between the two contrasting magmas, so the shape of the blobs of mafic magma tended to approach a spherical shape, thus reaching the smallest possible surface area relative to volume. Enclaves may contain phenocrysts of feldspar, and in the white granite these can be seen crossing the contacts of the enclaves (Fig. 6.10). Such large crystals are called '*dents de cheval*' (horse's teeth) (Read 1957; Pitcher 1993). They were regarded by transformists as evidence for the formation of granite by the solid state transformation of more mafic rock (Read 1957), but modern research involving spot-by-spot analyses of the crystals has shown that, on the contrary, different parts of the crystal were precipitated from magmas of different composition (Vernon 1986; Pitcher 1993). This is strong evidence against the view that large phenocrysts (megacrysts) in granite formed by metasomatic alteration after crystallization (e.g. Stone & Austin 1961). Enclaves are most frequently

Fig. 6.10. Two enclaves in white granite, Anglers Lane. Both host granite and enclave diorite are porphyritic, with a '*dent de cheval'* phenocryst growing across a contact.

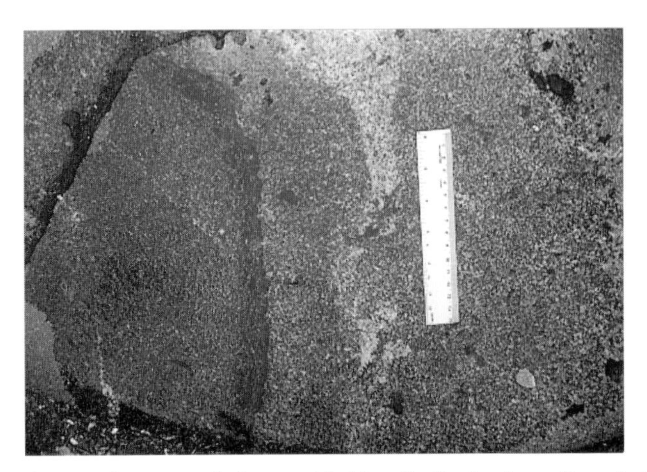

Fig. 6.11. Contact between darker and lighter diorite, by the railway bridge over Prince of Wales Road. The darker diorite (right) shows a chilled margin against the lighter diorite, which is cut by a vein of even lighter material, lacking a chilled margin.

found in the diorite kerbstones, confirming a conclusion of Didier & Barbarin (1991), based upon observation of large areas of well exposed granitic and dioritic rocks, such as are found in the Sierra Nevada batholith of California, USA. In diorite kerbstones it is also possible to see evidence for the complex and dynamic intermixing of more and fewer fluid magmas (Fig. 6.11), which Pitcher (1993) suggests is a major process leading to the formation of mafic enclaves.

Conclusion

The well polished kerbstones of Kentish Town allow naked eye and hand lens observation of heathen within granites and diorite, which confirm modern theories about the incorporation of country-rock fragments and the mixing of magmas during the intrusion of granitoid batholiths. Most importantly it illustrates the importance of the urban resource in demonstrating that petrology can be studied in cities as well as in the mountains, and that in some cases the rocks of our cityscapes may represent a valuable primary research data source.

Acknowledgements

I thank Eric Robinson for showing me the value of city petrology, and Alan Boyle and Hilary Downes for valuable discussion of modern research on granitoid magmas.

References

Anon. 1991. *The Times London History Atlas*. Times Books, London.

Barton, N. 1992. *The Lost Rivers of London*. Historical Publications, London.

Didier, J. & Barbarin, B. 1991. *Enclaves and granite petrology*. Elsevier, Amsterdam.

Middlemost, E. A. K. 1985. *Magmas and Magmatic rocks: an introduction to igneous petrology*. Longman, London.

Nockolds, S. R. 1932. The contaminated granite of Bibette Head, Alderney. *Geological Magazine*, **71**, 31–39.

Pitcher, W. S. 1993. *The Nature and Origin of Granite*. Blackie, Edinburgh.

Read, H. H. 1957. *The Granite Controversy*. Murby, London.
Stone, M. & Austin, G. C. 1961. The metasomatic origin of the potash feldspar megacrysts in the granites of Southwest England. *Journal of Geology*, **69**, 464-472.
Thomas, H. H. & Smith, W. C. 1932. Xenoliths of igneous origin in the Tregastel–Ploumanac'h Granite, Cotes du Nord, France. *Quarterly Journal of the Geological Society of London*, **88**, 274-296.
Tindall, G. 1977. *The field beneath: the history of one London village.* Maurice Temple Smith, London.
Trench, R. & Hillman, E. 1993. *London under London: a subterranean guide*. John Murray, London.
Vernon, R. H. 1984. Microgranitoid enclaves in granites-globules of hybrid magma quenched in a plutonic environment. *Nature,* **309**, 438-439.
---- 1986. K-feldspar megaliths in granites-phenocrysts, not porphyroblasts. *Earth Science Reviews*, **23**, 1–63.

7 The evolution of planning and use of urban parks and open spaces in Britain

Mike McGibbon

Summary

- The historical development of parks and open spaces is reviewed.
- Most urban parks date from the dramatic increase in urbanization in the nineteenth century.
- The development of these parks coincides with the heyday of natural history, and many parks contained geological components in their original design.
- Today parks and open spaces provide an opportunity for the introduction of geology into urban areas.

Parks and open spaces represent one of the most important resources of urban geology and geomorphology. These areas of our urban environment may be artificially constructed, marooned remnants of a natural landscape or by-products of industry, such as old quarries or railway cuttings. Despite its apparent richness, this resource is under-used and unsung. In order to appreciate the diversity and range of parks and open spaces in urban Britain, it is important to understand the nature of the process which created them. The purpose of this paper is to provide an overview of the evolution of parks and open spaces in British cities, with particular attention to London. It reviews the benefits derived from parks and open spaces by city residents, as well as examining the evolution of park design.

Park design developed from formal, laid out gardens with an idealized view of nature in the nineteenth century, through a focus on playing fields and physical recreation during the inter-war years, and eventually to parks as informally arranged natural spaces in which urban dwellers interact with nature and learn about the natural environment, and its ecology, geomorphology and geology.

The benefits of nature in the urban setting

Millward & Mostyn (1989) have illustrated that the benefits to be derived from nature clearly vary from one individual to another. They have demonstrated that three important factors may influence the perceived benefits that accrue to individuals. These are discussed below.

Attitudes and behaviour regarding leisure time and outdoor recreation

Research in the United Kingdom indicates that there is an increase in the amount of leisure time spent in, around or near the home, as opposed to special trips to other areas to satisfy recreational needs. The exception to this, however, is the daytrip to some special destination outside the immediate area, for example, the seaside, some historic site, wood, heathland or park. In terms of behaviour, people tend to regard natural areas or parks as multiple use areas. For example, a wood will often be used for jogging, kicking a ball around, cycling, picnicking and perhaps most importantly for urban dwellers, as a tranquil place to escape from the pressures of city life.

By far the most popular features of open spaces to visitors are the natural features of landscape (geology and geomorphology) and wildlife, peace and quiet, and the sense of freedom engendered (Walker & Duffield 1983). Most visitors engage in relatively passive activities such as walking or sitting enjoying the scene. Walking, as the most common activity, may be affected by the opportunity to play sport, for example an open green space for football or cricket, or by the presence of a car park with a scenic view, in which case many people simply sit in the car and look at the view.

Perceptions of urban wildlife and landscape

People tend to perceive landscapes positively if they include the following features: lots of mature trees - parks are rated as giving a high degree of satisfaction if they are dominated by trees; lush vegetation - the more vegetation, grasses and meadowland present the more positive the response of the users (Ulrich 1981); naturalness - the landscape that is perceived as having evolved naturally is regarded more highly than one perceived as being planned or designed; textural variety of the ground, that is some relatively manicured parts along with relatively wild parts is regarded as positive, as long as it is not too rough or scruffy; sensitive use of features such as benches, sheds, tables which fit in with the general environment; the presence of water such as ponds and streams; and the presence of bird life.

Personal satisfactions derived from nature

Instinctively we accept the idea that nature is good for us. However, social researchers have tried to understand more fully what being with nature actually does for people emotionally, intellectually, socially and physically. Studies based on what respondents report as benefits of being with nature have generally found that given a free choice people will choose a park or open space with natural features (geomorphology, geology, wildlife, etc.), although they often have difficulty explaining why. Using a variety of techniques including role play, drawing, indirect questioning and careful self-assessment of feelings, researchers have found that there are four types of personal benefits which people obtain from nature: emotional, physical, intellectual and social benefits.

The emotional benefits for urban dwellers include escape from the city, often expressed as getting away from the concrete jungle, or relaxing in nature, or finding a peaceful hideaway, a place to unwind and think. Even the presence of a natural park is reassuring to urban residents who only use it occasionally.

The physical benefits include the exertion of walking in the park, stimulating the whole body and all of the senses: the sounds of birds and the wind in the trees; the smell of flowers and shrubs; the sight of the variety of colours and textures of landforms, exposed geology, grasses, trees and shrubs; the feel of vegetation and the fresh air.

The intellectual benefits include an awareness of being an integral part of nature, thereby enhancing one's self-identity. There is nowadays a general fascination about the way nature works, how the change in the seasons is demonstrated in natural life cycles. There is the benefit of seeing trees, flowers, plants and animals and their interaction with landscape other than in a book.

The social benefits are said to include the perception that one is in a country setting where people are more friendly to one another. As a result, country friendliness is more likely to be exhibited in people's behaviour. Milward & Mostyn (1980) suggests that people walking in the park, for example, are much more likely to interact in a small town, friendly fashion than outside the park.

Over time, attention has shifted from some of these benefits to others, depending on changes in the meaning society attaches to the terms recreation and environment. The development and evolution of urban parks since the nineteenth century also reflect such changes in perceived benefits and meaning of recreation and environment. Accordingly, the following section of the paper reviews the evolution of urban parks within this context.

The evolution of British urban parks

British cities since the mid nineteenth century have been renowned for the provision of urban park space. The creation of the municipal park has been seen as a prime example of the Victorian passion for reform, designed to improve the physical, moral and spiritual condition of the urban dweller (Chadwick 1966). In our cities we have a variety of open park spaces. In London, these include the Royal Parks such as Hyde Park, St James's Park, Green Park, Greenwich Park and Regent's Park, designed by John Nash between 1811 and 1826. However, the general public did not necessarily have the right of access to such parks. Municipal parks, within the control of the local authority, and including the right of public access for the first time, were created largely as a result of the passage of the Municipal Corporations Act of 1835, which created the first formal structure for proper local government in British cities. The first municipal park, Moor Park in Preston, predated that legislation by two years. The need for municipal parks grew from the dramatic urbanization of the eighteenth and nineteenth centuries. By 1801, Manchester, Liverpool, Birmingham and Bristol all had populations of over 60 000 and the population of London was well over one million. By 1851, the populations of Manchester and Liverpool had reached nearly 400 000 each and Birmingham's population was well over 200 000. Until that time there had not been any enormous need to set aside open space for recreation or conservation of nature. Overcrowding, insanitary conditions and inadequate public facilities led to low life-expectancy and high infant mortality. Epidemics of cholera, typhoid and typhus killed many thousands. The pressure to provide parks stemmed, at least in part, from the belief that the air itself caused illness and that parks and open spaces would act, in William Pitt's words, as 'lungs for the city'; the green ventilators (Welsh 1991).

The Government's 1833 Select Committee on Public Walks (SCPW) reported that there was a great shortage of public open space for the use of the working classes of the large industrial cities. Much of the open space that did exist, such as some of the early botanical gardens, was open only to subscribers and closed to everyone else. Nonetheless, the SCPW report marked the beginning of a period during which social reformers lobbied parliament for legislation to establish walks, parks and playgrounds. These efforts led to the creation of Primrose Hill Park (1842) in Central London, and Victoria Park (1842) in Tower Hamlets to serve the needs of the industrial workers of the East End. In the period between 1842 and 1859, local authorities throughout the country acquired parks by a variety of means. In some cases they benefited from the generosity of philanthropic industrialists like Sir Titus Salt, a textiles magnate who donated money for the creation of Peel Park in Bradford and who persuaded other industrialists to do the same. In other cases, local authorities were able to obtain a central government grant from a £10 000 fund, established in 1841.

Most of the early parks were in the industrial towns of northwest England, reflecting an association with the major industries of the time and the recommendations of the SCPW. Government legislation contributed to the emergence of municipal parks, through the 1847 Towns Improvement Clauses Act which simplified procedures by which local authorities could acquire land for parks. The 1848 Public Health Act gave Local Boards of Health the power to provide, maintain and improve land for municipal parks.

Between 1845 and 1875 park development increased steadily, but it was after 1875 that a substantial increase in activity became evident as a result of the Public Health Act of that year. This was the first major statutory provision that enabled local authorities to acquire and maintain land for recreation. This act was especially important because it made provision for loans to local authorities to develop parks. In 1875 three applications for loans were made. By 1890, the number of applications had risen to 25 and some local authorities were making several applications in one year. Land acquired under this act could not be used for purposes other than parks or recreation.

The resulting parks were formally planned and laid out with walks and views carefully selected to provide natural vistas and visual variety for the user. The aim was to bring elements of the idealized countryside into the city. Natural history was exceedingly popular during this time and as a result the greatest development of the potential of geology and geomorphology in these parks was made. Artificial landforms, such as the reconstructed model of the waterfall Thornton Force in Lister Park in Bradford, or the artificial rockwork cascades in Battersea (Doyle *et al.* 1996) were created in sympathy with the fashion for geology. Many park walks were embellished with plants and flowers brought back to Britain from around the world by naturalists and explorers. This approach was extremely popular with the public. One of the most popular was Crystal Palace Park at Sydenham, designed by Sir Joseph Paxton to include formal gardens, winding paths, large-scale tree planting, extensive use of flowering plants as a public attraction and, uniquely, an integrated 'geological theme park' with constructed displays of geology with dinosaurs and other extinct animals in 'natural' settings, illustrating the excitement generated by the work of Darwin and other scientists of the day (Doyle & Robinson 1993; Doyle *et al.* 1996). The natural walks created were carefully separated from play areas such as archery grounds, football pitches, swings, skittle alleys, cricket grounds and the like (Fig. 7.1).

Together, these design elements reflect a concern with the quality of the residential environment in terms of opportunities for healthy recreation, a reminder of the aesthetic value of the countryside and of nature, an opportunity to improve the physical and moral condition of the population, and its education. The idea of education as part of the concept of recreation, is well illustrated by the inclusion of complex geological tableaux in some parks, and by the fact that museums or free libraries were sometimes included in parks because they were regarded as conducive to betterment of the population. In general, the urban park can be interpreted as reflecting a desire to preserve a pre-industrial past in the face of the social, economic and environmental degradation produced in the emerging industrial cities (Newby 1990).

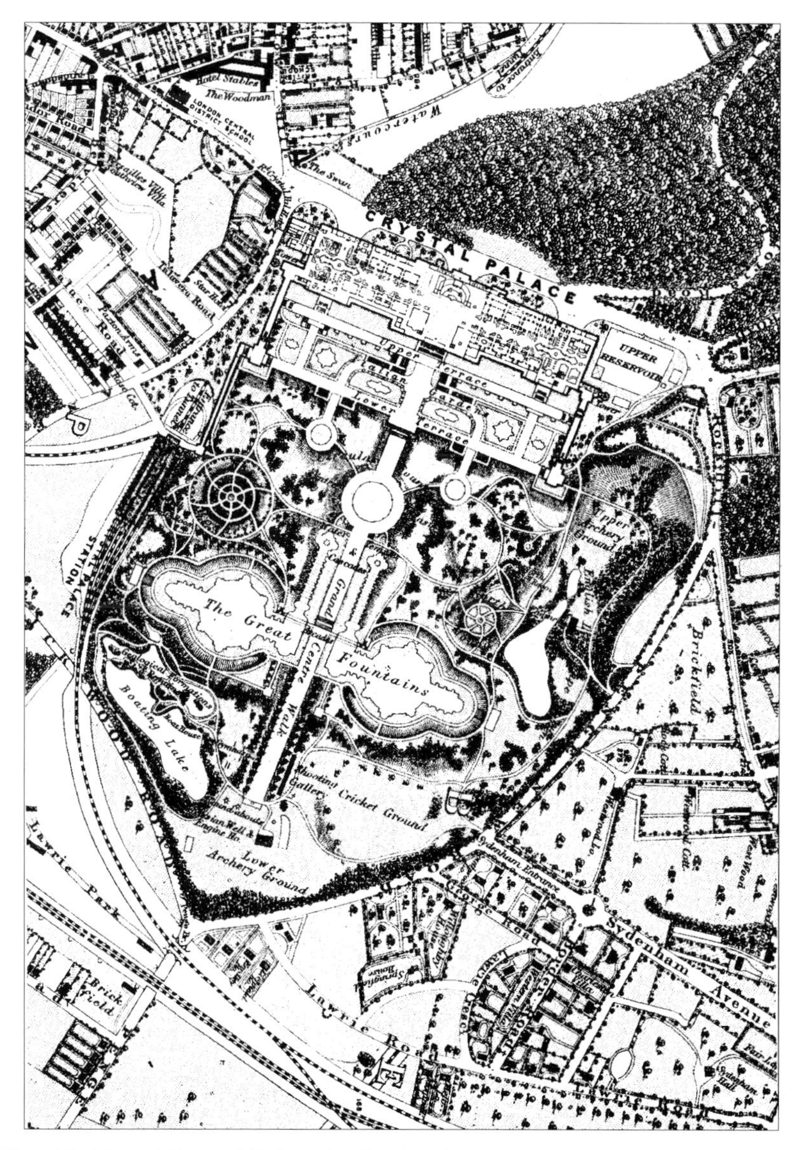

Fig. 7.1. Map of Crystal Palace Park, showing the space set aside for different activities.

The development of urban parks and open spaces continued in a similar vein until the 1930s when the emphasis turned to active recreation as part of the government's drive to produce a population fit for war. In 1937, a national fitness campaign was instituted and local authorities were encouraged to build playing fields for football, rugby, cricket and hockey. These later came to be criticized as bleak, green deserts by conservationists (Welsh 1991).

The concept of the green belt

Two influential planners, Raymond Unwin and Patrick Abercrombie, followed the lead set by Ebenezer Howard (Howard 1902) at the turn of the century and began to address the relationship between the city and the countryside. Along with the increasing scale of urbanization went

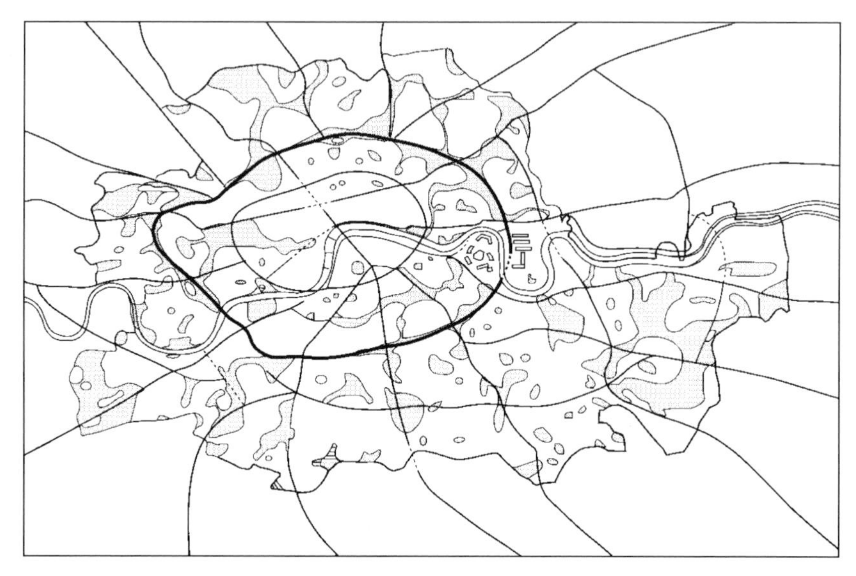

Fig. 7.2. Abercrombie's proposed system of open spaces within London. (Modified from Forshaw & Abercombie 1943).

increasing concern, both moral and aesthetic, for the countryside which was being swallowed up. Official open space planning began with the First Report of the Greater London Regional Planning Committee (1929). In this, Unwin recommended that a 'Green Girdle' be created around London for playing fields and recreation. Subsequently, the Greater London Plan produced by Abercrombie (LCC 1945) included what Turner (1992) calls 'one of the most brilliant open space plans ever prepared for a capital city'. Abercrombie argued for the clear separation of town and country in order to preserve the country landscape and to retain nature as a preserve from modern urban life (Healey & Shaw 1994). His aim was to create a coordinated park system for the entire Greater London region (Fig. 7.2). All forms of open space were to be considered as a whole and would be coordinated into a closely linked park system, with parkways along existing roads forming the links between the larger parks.

The green belt proposals in the 1944 Greater London Plan made a distinction between a green belt ring (about 8 km deep) and an outer country ring (Fig. 7.3). Recreation was to be the predominant use of the green belt ring, and agriculture and mineral extraction the prodominant uses in the outer country ring. No new buildings were to be erected in

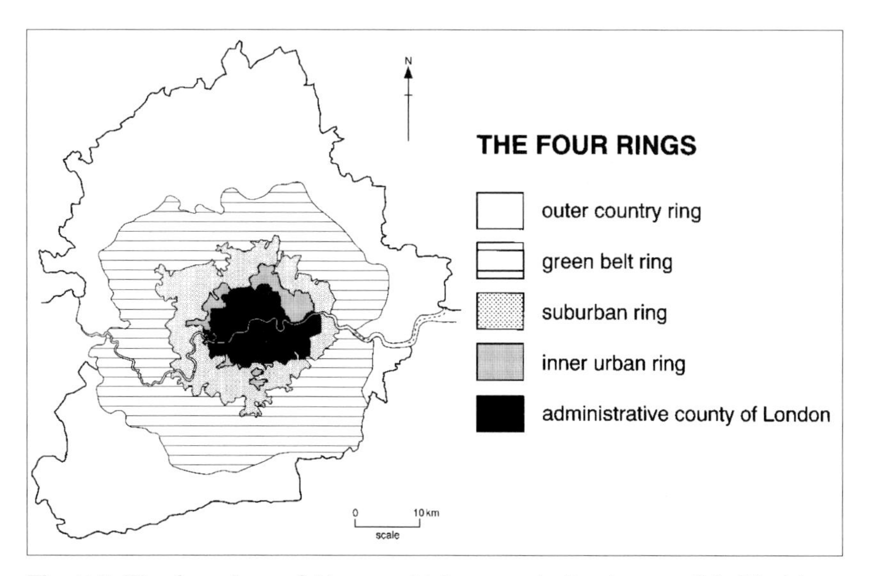

Fig. 7.3. The four rings of Abercrombie's green belt scheme. (Modified from Abercombie 1943.)

either ring. The necessary control of building development to achieve this was provided by the Town and Country Planning Act of 1947. One of the most important aspects of the green belt to the earth sciences was that it meant that there was potential for studying geology and geomorphology in its natural setting. Landforms such as rivers and valleys for example, are particularly valuable components of the green belt as they both enhance the naturalness of the setting and provide real geomorphological features for appreciation and study.

The era of growth management

The London Government Act of 1963 established the Greater London Council (GLC) as a strategic planning authority with a responsibility for an area extending to *c*. 22 km from central London. In 1969 a draft Greater London Development Plan (GLDP) was published, and finally approved in 1976. The open space provisions in the plan centred on three concepts: the Green Belt; Metropolitan Open Land and Open Spaces. The Green Belt occupied 90 000 acres or 36 423 hectares within the GLC area. The concept of Metropolitan Open Land was introduced as a protective designation for open spaces within the urban area. It was recommended that parks, woodlands, golf courses, nursery gardens, cemeteries and other open land which might be developed should receive this designation. This is particularly important for urban geology as areas such as cemeteries are a timeless resource for the study of rock type and weathering (Robinson 1996).

The Open Spaces category meant public parks. An assessment of the provision of public parks was carried out based on an 'open space hierarchy', and by that measure a number of quite large areas was identified as being deficient in metropolitan parks. However, virtually nothing was done about most of these deficient areas (Turner 1992). Action was recommended on only two sites: one south of the Thames and one north. In south London, Crystal Palace Park was to be extended and more facilities provided there. In north London, improvements were to focus on what was called the Dagenham Corridor. There, the deficiency was to be made up by linking or extending two existing large parks.

The draft plan was subjected to a rigorous inquiry. The plan was criticized for being half-hearted in its commitment to the recreational uses of the green belt and lacking firm statements directing that local plans must provide for the rehabilitation of derelict land. The concept of Metropolitan Open Land was criticized. It was recommended that detached patches of open land or fringes of the green belt should be regarded instead as part of the Green Belt. The final draft of the GLDP was approved in July 1976. In it a stronger protective statement about the Green Belt was included. The concept of Metropolitan Open Land was included as land which needed to be protected as much as the Green Belt.

Era of ecology as a focus of interest and debate

One important gap in the 1976 plan was the absence of any direct reference to nature and earth science conservation. That ommission was addressed in 1983 when a series of policies was created which established conservation priorities and management requirements for conservation sites and species throughout Greater London. In the main, they were not approved as alterations to the plan, but a London Ecology Centre was set up which, in practice, had a great deal of influence on open space planning. Every borough now has policies for nature and earth science conservation in its local plans.

Since the abolition of the GLC in 1986 London has had no strategic planning authority. The London Planning Advisory Council (LPAC) was formed to fill the vacuum. Since 1988/89 its role has been to give 'strategic planning advice' on London planning to the Department of the Environment (DOE). The DOE provides 'strategic guidance' to the 33 local planning authorities in the former GLC area. Both agreed that the Green Belt must be permanent. One of the concepts highlighted in both guidance and advice was the concept of 'Green Chains' - interlinking of open spaces, footpaths, rivers, canals, bridleways and disused railways for recreation and nature conservation purposes. An example is in southeast London (Fig.7. 4) where the green spaces have been designated as Metropolitan Open Land. Importantly, the Green Chain Walk links areas of considerable geological and geomorphological significance: Crystal Palace, Greenwich Park, Gilberts' Pit close to Maryon Park in Charlton, a geological SSSI, Abbey Wood, also a geological SSSI and one in which fossils may be collected. Links are provided by a signposted

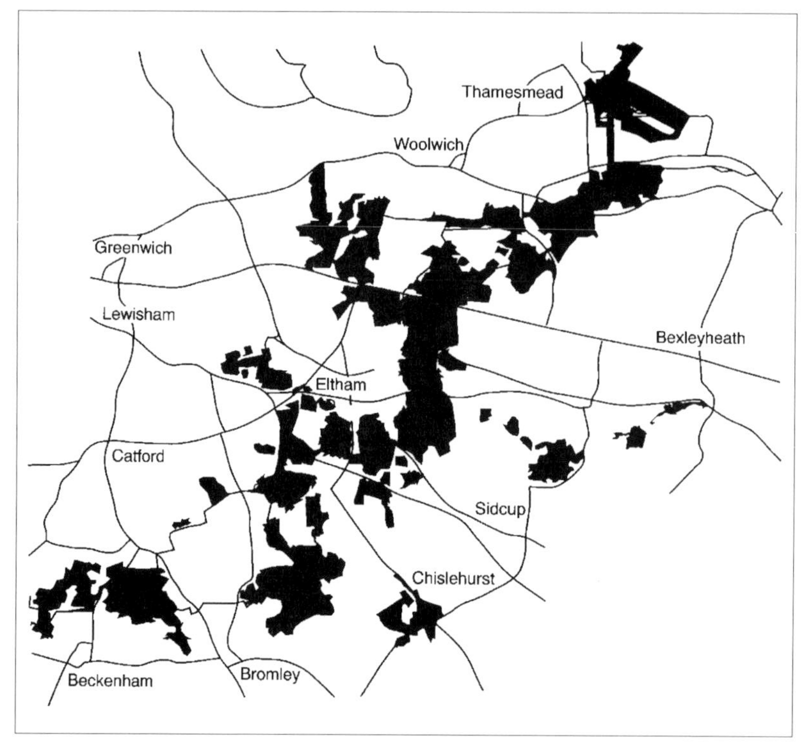

Fig. 7.4. Southeast London Green Chain. (Modified from Turner (1992).)

walkway through the urban areas which interrupt the flow of open space, and interpretative guide leaflets are provided in libraries and municipal offices.

Other open space links are being developed in London including riverside walks, countryside walks, civic walks and canalside walks. In addition, great emphasis is now placed on nature and earth science conservation. Strips of semi-natural landscape beside railways, rivers and canals are now very desirable as recreational spaces which educate an increasingly interested general public about the geology and ecology of their city. These are seen as natural corridors and may become part of a London-wide web of wildlife habitats.

The popular image of most urban space in cities is of the traditional park landscape originating in the nineteenth century municipal parks movement, or the sports fields which were so popular from the 1930s, that is, one of mown grass, scatterings of trees and shrubs, a standard set of play equipment for children, and pitches and courts for organized sport. At the present time, however, some regard such use of open space as a reflection of the failure by planners to recognize the full significance of open space to life in the city. Alternatives to such formally planned open spaces are the numerous so-called 'natural' open spaces to be found in the cities. These are particularly significant in terms of urban geological resource, as apart from disused railway lines, they include abandoned quarries, woods, areas of subsidence and river valleys, especially those susceptible to regular flooding. The key characteristic of these natural open spaces is that they are virtually, if not totally, unmanaged. This has tended to appeal to conservationists who have found them to be rich in geological outcrop, natural landforms and wildlife and able to provide urban dwellers with an opportunity to observe the natural world close to their home rather than in the countryside.

The main influence on the creation of urban geology and ecological areas in Britain seems to have come from the fields of nature conservation and landscape design. Only more recently have urban planners become important in the development of urban nature conservation. Throughout the 1960s and 1970s, the debate about the human relationship with the natural world and the exploitation of natural resources created a momentum for change to sustainable use of natural resources (Healey & Shaw 1984). In the field of nature conservation, that has been translated into the emergence of various urban ecology organizations and urban wildlife groups throughout the country. Many of these wildlife groups have geology and geomorphology on their agenda now through the development of the RIGS (Regionally Important Geological Sites) initiative by the former Nature Conservancy Council and The Wildlife Trusts. The priorities for these organizations have been: education, recreation, and aesthetics. Common themes are: (1) providing city dwellers with the opportunities to experience natural landscapes within towns and cities; (2) improving awareness among land use decision makers of the natural conservation value of their land and of the benefits to be had from adopting conservation principles in the management of that land; and (3) providing educational programmes for children to learn about the natural world and to develop a sense of responsibility towards it.

Conclusion

Since the development of the parks movement in the nineteenth century open spaces have been considered beneficial to the populace, as 'green ventilators' to ward off disease, as educators and improvers of the mind, and as areas where recreational pursuits can be carried out. Sociological research certainly shows that parks and open spaces are extremely important to the local populace and visitor alike for a variety of reasons. The concept of 'naturalness' is extremely important and has been since the development of parks in the romantic landscape tradition through to the present day ideals of nature conservation. In all of these, a natural landscape of landforms and rocks is an integral part and has been since the Victorian innovation of recreating totally artificial rockworks through to the development of Green Chains linking geological sites in an urban environment. In all these cases, the development of parks and open spaces should be seen as a significant achievement by planners in providing the urban environment with one of its most important geological resources.

References

Abercombie, 1943. *Greater London Plan 1944*. London.

Chadwick, G.F. 1966. *The Park and the Town*. Architectural Press, London.

Conway, H. 1991. *People's Parks, the design and development of Victorian Parks in Britain.* Cambridge University Press, Cambridge.

Doyle, P. & Robinson, E. 1993. The Victorian 'geological illustrations' of Crystal Palace Park. *Proceedings of the Geologists' Association*, **104**, 181-194.

Bennett, M. R. & Robinson, E. 1996. Creating Urban Geology; a record of Victorian innovation in park design. *This volume.*

Forshaw, & Abercombie, 1943. *Country of London Plan.* Macmillan, London.

GLRPC 1929. *First Report of the Greater London Regional Planning Committee.* London.

Healey, P. & Shaw, T. 1994. Changing Meanings of 'Environment' in the British Planning System. *Transactions of the Institute of British Geographers,* **19**, 425-438.

Howard, E. 1902. *Garden Cities of Tomorrow.* (Reprinted 1951) Faber and Faber, London.
London County Council. 1945. *Greater London Plan 1944*. HMSO.
Milward, A. & Mostyn, B. J. 1980. *Personal benefits and satisfactions derived from participation in urban wildlife projects. Shrewsbury.* Nature Conservancy Council Interpretive Branch.
---- & ---- 1989. *Urban Wildlife Now; people and nature in cities*. London. Urban Wildlife Now Series, **2**.
Newby, H. 1990. Ecology, Amenity and Society; social science and environmental change. *Town Planning Review*, **61**, 3-13.
Robinson, E. 1996. 'The Paths of Glory...' *This volume.*
Turner, T. 1992. Open Space Planning in London; from standards per 1000 to green strategy. *Town Planning Review*, **63**, 365-386.
Ulrich, R. S. 1981. Natural versus urban scenes. *Environment and Behaviour,* **13**, 523-556.
Walker, S. E. & Duffield, B. S. 1983. Urban Parks and Open Spaces - an overview. *Landscape Research*, **8**, 2-12.
Welsh, D. 1991. *The Management of Urban Parks.* Longman, London.

8 Creating urban geology: a record of Victorian innovation in park design

Peter Doyle, Matthew R. Bennett & Eric Robinson

Summary

- Parks are an important geological resource in urban areas.
- Many large municipal parks contain some form of 'geological illustration', a product of the Victorian enthusiasm for natural history.
- The Victorian geological exhibits of Crystal Palace Park and Battersea Park are amongst the best examples of their type, and these are discussed.
- Finally, the question is posed: if the Victorians could incorporate meaningful artificial geology into the urban open spaces they designed, why is it seemingly impossible today?

The heyday of natural history is unequivocally the Victorian era. Concomitant with the development of the industrial wealth of Britain was a desire to entertain through 'rational' amusement intended to stimulate the mind rather than the baser instincts (Barber 1980). Natural history became a national mania, with the desire for examining the minutiae of animal and vegetable objects through microscopy, for the development of aquaria and 'Wardian cases' for ferns, and in the appreciation of the new science of geology and its sister subject, palaeontology (Barber 1980). Undoubtedly, a part of this developing interest in natural history was the creation of the municipal park in the major industrial towns and cities of Britain (Conway 1991; McGibbon 1996). In many cases the function of these parks was extended beyond a simple desire to demonstrate municipal standing, or to provide an open space within an industrial town or city, as it is clear that the desire to educate was influential, at least in part, in the final development of the park. In many cases, it is clear that landform and geological features were as important to the landscape designers as the appreciation of the wildlife and vegetation, and some innovators strove to include geological or landform features.

Some parks, such as Victoria Park in Glasgow, benefitted from the occurrence of natural geological features, in this case the famous fossil grove of Carboniferous lycopod tree stumps which has been conserved *in situ* since the late nineteenth century and is one of the earliest examples of earth science conservation in existence. Most, however, were bereft of naturally occurring geological features, and designers had the opportunity to create for the benefit of the visiting public the unique resource of an artificially created landscape. A good example is found in Birkenhead Park, opened in 1847. Reputedly the first park to be provided from public money, Birkenhead Park was the dream of Sir William Jackson, one of the towns first commissioners, and was designed by Sir Joseph Paxton. The park was laid out in the naturalistic style of the time and was built upon marshy barren ground (McInniss 1984; Thornton undated). It was constructed in two parts, divided by a roadway, and had as its main theme two lakes surrounded by earth mounds, created from the spoil dug from the lake basins. The general impression created was that of a natural landscape of a river meandering in its valley. A large rockery was created using coarse Pennine gritstone blocks; although this is visually impressive, little regard was given to geological accuracy, and the overall impression is not one of a naturally occurring outcrop but rather a coarse scree slope (Fig. 8.1).

If Birkenhead Park is typical of the generally sympathetic view towards geology and geomorphology in the parks which were being created in Britain during the late Victorian era, then two parks stand out as excellent examples of further development of geology in harmony with landscape: Crystal Palace Park and Battersea Park, both in south London. In these parks, complex geological relationships were reconstructed in a completely artificial way, innovative enough to inspire interest in naturally occurring geological materials. In London, few parks can draw upon naturally occurring geological or geomorphological features. Most natural geological features occur in larger open spaces, the so-called lungs of London, such as Hampstead Heath or Epping Forest. Geological Sites of Special Scientific Interest (SSSI) with natural outcrop are found in the urban environment, such as Gilberts' Pit SSSI in Charlton or Harefield SSSI in Uxbridge, which expose important Palaeogene sediments. Both are remnants of aggregate and other bulk mineral extraction and are now conserved as features of enclosed spaces, rather than formally contained within a park infrastructure. A rare example of a Royal Park in London

Fig. 8.1. Rock crags in Birkenhead Park; a contemporary postcard view from 1907.

with a natural geomorphological component is that of Greenwich Park. Here, a steep slope formed by Palaeogene sediments grades down from the flat expanse of Blackheath (formed on the well drained Blackheath beds) to the Thames river terrace area close to Greenwich town, giving the park a dramatic landscape appeal that is unique amongst London's parks.

Battersea and Crystal Palace are important in illustrating that geology can be successfully demonstrated using what are otherwise 'alien' materials in a London Clay geological setting with little hope of outcrop or dramatic landscape feature. This is something which is not restricted to parks in London, as Lister Park in Bradford contains a reconstruction of the angular unconformity at Thornton Force (Fig. 8.2), and other examples probably exist. The purpose of this paper is not to duplicate the work that has already been carried out in documenting the development and rationale behind the unique Victorian geological reconstructions at Battersea and Crystal Palace (Doyle 1993, 1995; Doyle & Robinson 1993, 1995; Robinson 1994), but rather to demonstrate through a pictorial record the validity of the Victorian approach to geological reconstruction in the present day park and open space environment.

Fig. 8.2. The sorry state of the reconstruction of the Yorkshire waterfall, Thornton Force, built in 1902 in Lister Park, Bradford. This demonstrates the famous unconformity. Visible at the bottom of the picture are vertical Lower Palaeozoic slates overlain by horizontal Carboniferous Limestone.

Crystal Palace Park

Crystal Palace Park is justly famous as the home of the first full-scale reconstructions of dinosaurs and other extinct animals anywhere in the world, constructed in 1853-54. The story behind the development of these animals is discussed in Doyle & Robinson (1993, 1995) and McCarthy & Gilbert (1994) and is not discussed further here. However, the integrated nature and complexity of the scheme, involving not just the dinosaurs and other animals, but appropriate rocks in stratigraphical order and exhibiting a complexity of geological structure, has until recently remained forgotten (Doyle & Robinson 1993).

The geological exhibit was designed by Professor Ansted, then Secretary of the Geological Society, in conjunction with Sir Joseph Paxton, the originator and designer of the park as a whole. Benjamen Waterhouse Hawkins, a scientific illustrator and modeller, created the extinct vertebrate animals, guided by Sir Richard Owen.

The scheme was intended to make sound geological sense. Dipping Palaeozoic rocks were overlain unconformably by Mesozoic and Cenozoic rocks. The economic resources of Britain were represented in their 'natural' state: coal, ironstone and lead mineralization are all present.

There is a complexity of faults, some obvious, some to be inferred by the astute observer. The extinct animals all rest on the rock types which contained their fossil remains. The abbreviated stratigraphical column of England is present.

Elements of the original scheme, forgotten or ignored from late Victorian times, have now been reintroduced to the public using a trail guide approach (Doyle 1995). The local authority, the London Borough of Bromley, in partnership with the Sports Council (who operate the National Sports Centre in the middle of the park), a local history group, the Crystal Palace Foundation, and the University of Greenwich intend to restore the exhibit to some part of its former glory.

The following illustrations demonstrate the relative complexity and sophistication of the exhibit, achieved in an artificial manner within a public park (Figs 8.3-8.6). Figure 8.3 displays some of the complexity of the Palaeozoic exhibit: it has beds of ironstone in continuous and nodular horizons, a coal seam and sandstone bed. The intervening 'beds' were originally a cement rendering with delicately sculpted cross-beds and other accurate features. Two faults can be seen, and the whole presents an impression of an accurate model of the geology of the Clay Cross Coal Measures in Derbyshire. This was balanced in the original exhibit by the Carboniferous Limestone cave and lead mine. Only the cave remains (Fig. 8.4). The park is most famous for its reconstructed animals: two are illustrated here. The dinosaur *Iguanodon* from the Lower

Fig. 8.3. The reconstructed Coal Measures in Crystal Palace Park, London. This is based on Clay Cross in Derbyshire. The blocks of stone used in this reconstruction, including the coal, were transported from Derbyshire.

Fig. 8.4. Interior of a reconstructed limestone cave at Crystal Palace Park, London. It shows stalactites and stalagmites, and formed part of a lead mine within the Carboniferous Limestone exhibit.

Fig. 8.5. The *Iguanodon* standing upon Wealden Sandstone in Crystal Palace Park, London.

Cretaceous of the Weald is reconstructed on a rock 'outcrop' of Wealden sandstone (Fig. 8.5). The giant ground sloth *Megatherium* (Fig. 8.5) is commonly encountered in the Quaternary gravels of South America. It is reconstructed in a dynamic pose, resting on a base of coarse aggregates representing its host geology (Fig. 8.6).

Fig. 8.6. The *Megatherium* standing upon reconstructed Quaternary gravels in Crystal Palace Park, London.

Battersea Park

Created from reclaimed land (Robinson 1996), Battersea Park has even less natural relief than Crystal Palace, which is developed upon a gently sloping site. The park itself is presently being restored by the London Borough of Wandsworth and central to it is a 15 acre lake with several headlands and embayments. The public is allowed to access these areas which give an impression of walking around an island site. The most impressive component of this site are the completely artificial 'rock crags' which dominate the southern end of the lake. Approaching from the Main Avenue or the Chelsea Bridge Gate, the eye is caught by these rocky crags which rise opposite the boat landing stage. The crags are now particularly prominent in that they are crossed by a cascade of flowing water which forms a striking focus to the scene. You need to skirt around the lake at close quarters to be able to appreciate that the complex cascades are fashioned from artificial stone, skillfully copying features such as bedding and jointing so as to complete the deception.

The 'crags' at Battersea were the work of James Pulham, a landscape designer who was very active at the time of the laying out of the Park in the years between 1866-70. His work otherwise was to help landowners to improve the parklands which surrounded their country seats. Using the best available cements and bulk sand of the day, Pulham even went so far as to mimic graded bedding such as we would expect to see in the Millstone Grit of the Pennines, for example. Sadly, the work was done as a veneer upon a rubble base, using a concrete-based material usually referred to as 'Pulhamite', and this has been revealed by the wear-and-tear of time. Recent restoration of such damage has been undertaken by English Heritage, who help to maintain the buildings of the Park, but no one would be deceived by what has been done. It seems that the skill of Mr Pulham and his artificial stone died with the last of the family in business. The detail of the construction of these crags has been described by Elliott (1984) and Robinson (1994), and a comprehensive review of known Pulhamite crags, including Battersea, is given by Festing (1984).

The Battersea crags are not comparable with those 'exposed' in Crystal Palace in that they do not use natural materials other than the aggregates in the concrete mixture. However, the effect is dramatic and demonstrates both geomorphological and geological features with a surprising degree of complexity.

The following illustrations demonstrate the complexity and intricacy of detail in the exhibit (Figs 8.7-8.10). Figure 8.7 conveys the impression of the Pulhamite rockworks, but they have been heavily 'restored' with a concrete-based coating. The rockworks are impressive as there is the sense of alternating beds of coarse and fine sandstone, with a concordant

Fig. 8.7. The Pulhamite rock-works in Battersea Park, London. Although heavily restored these crags still display an accurate impression of real geology where no natural exposure existed.

Fig. 8.8. Restored Pulhamite rock-works at Battersea Park, London. The rock-works have been restored using 'guncrete', which has altered their original appearance. This can be demonstrated by comparing this photograph with Figs 8.9 and 8.10.

Fig. 8.9. Unrestored Pulhamite rock-works at Battersea Park, London. The rock-works were constructed using brickwork and York Stone slabs. Coated with Pulhamite the visual impact is that of a real sandstone.

dip. Even to the casual observer this appears accurate because it is in harmony with naturally occurring features in the Pennines, for example. Figure 8.8 shows the recently restored Victorian water cascade but shows the loss of detail to the rockworks, striking when compared with Figs 8.9 and 8.10. These figures show an unrestored part of the rockworks. They show an astonishing attention to detail. The thick beds were created by a coating of Pulhamite over blockwork; the thin beds, slabs of York Stone,

Fig. 8.10. Detail of unrestored Pulhamite rock-works at Battersea Park, London. This photography shows 'beds' of massive sandstone separated by fissile York Stone.

perhaps even paving slabs, used to demonstrate a change in geological texture. The cave may have been used to house live birds, but gives a good impression of a natural outcrop of sandstones. The angle of dip on these beds is concordant with the main restored cascade.

Conclusion

The development of the parks led to the desire in some of them to create or recreate some element of the natural world within the urban townscape. Both Crystal Palace and Battersea are supreme examples of Victorian ingenuity in recreating from almost nothing the original feel of naturally occurring geology and geomorphology. They demonstrate clearly the need for innovation in representing the natural world - geology and geomorphology - sympathetically in the context of the urban environment. They also demonstrate that the use of artificial materials need not mean that the exhibit has no geological accuracy or relevance. On the contrary, the full educational resource of both exhibits remains to be tapped by the modern educational establishment. If the Victorians could incorporate meaningful artificial geology into the urban open spaces they designed, why is it seemingly impossible to do so today? We have much to relearn from our Victorian forebears, the real innovators in urban geology.

References

Barber, L. 1980. *The heyday of natural history*. Jonathon Cape, London.
Conway, H. 1991. *The people's parks*. Cambridge University Press, Cambridge.
Doyle, P. 1993. The lessons of Crystal Palace. *Geology Today*, **9**, 107-109.
---- 1995. *The geological time trail. Paxton's heritage trail guide 2*. London Borough of Bromley.
---- & Robinson, J. E. 1993. The Victorian 'geological illustrations' of Crystal Palace Park. *Proceedings of the Geologists' Association*, **104**, 181-194.
---- & ---- 1995. Report of a field meeting to Crystal Palace Park and West Norwood Cemetery, 11 December 1993. *Proceedings of the Geologists' Association,* **106**, 71-78.
Elliott, B. 1984. We must have the noble cliff. *Country Life*, **175**, 30-31.
Festing, S. 1984. Pulham has done his work well. *Garden History*, **12**, 138-158.
McCarthy, S. & Gilbert, M. 1994. *The Crystal Palace dinosaurs*. Crystal Palace Foundation, London.
McInniss, J. 1984. *Birkenhead Park*. Countywise, Birkenhead.
McGibbon, M. 1996. The evolution of planning and use of urban parks and open spaces in Britain. *This volume*
Robinson, J.E. 1994. The mystery of pulhamite and an outcrop in Battersea Park. *Proceedings of the Geologists' Association,* **105**, 141-143.
---- 1996. A version of 'The Wall Game' in Battersea Park. *This volume*.
Thorton, C.E. Undated. *The people's garden*. Williamson Art Gallery, Birkenhead.

9 Geomorphological conservation: opportunities afforded in Greater Bristol

Eileen J. Pounder

Summary

- The importance of Regionally Important Geological/Geomorphological Sites (RIGS) is discussed with reference to the Greater Bristol areas.
- The public perception of three urban geomorphological sites is examined via a simple questionnaire
- The results demonstrate a clear interest on the part of the general public in the landscape around them and a wish to see it conserved.
- Education, public awareness and local planning are the keys to the effective conservation of such sites.

The Greater Bristol Conservation Strategy (NCC 1991) was brought together through cooperation between local authorities, the Bristol Development Corporation and conservation organisations under the guidance of English Nature (the Nature Conservancy Council, NCC, at the time). Wildlife conservation opportunities were reviewed and although geology featured in this review, little mention was made of geomorphology. Earth science conservation should encompass geomorphology (NCC 1990) and its omission from the document appears to be by default rather than by policy. However, the rationale stated in the strategy that 'the more people become personally involved with a site, the more likely they are to cherish and thus help to protect it', holds for geomorphological sites as well as any others. The difficulty may lie in the lack of general knowledge about geomorphology on the part of the public and the fact that the subject is usually in the remit of geography rather than geology.

This paper reports on some of the developments which have taken place in Bristol with respect to geology and focuses on the potential for geomorphological conservation and interpretation. The associated problems and opportunities are identified, a claim is made that the general public are interested in geomorphological features and a strategy for developing the potential which the subject offers in relation to urban geological conservation in Bristol is presented.

Developments to date

Significant work has been done by the Bristol/Avon Regional Environmental Records Centre and Bristol City Museum in site documentation. The Museum has done much to further geological education (Mathieson 1980). Bristol, for an urban area, has a remarkable amount of visible geology and geomorphology on its doorstep.

Site recording in the 1970s, as part of the National Scheme for Geological Site Documentation, provided the database for the Bristol/Avon Regional Environmental Records Centre. This has formed the basis for identifying the 187 Regionally Important Geological/Geomorphological Sites (RIGS) in Avon, of which 40 are within the urban or suburban area of the city. All the sites were visited and described by a team funded by the Manpower Services Commission and collated by Andrea Selby in 1986. This was updated in 1987 by Roger Vaughan and documented by John Hart in 1990. This provided the database for information used in the Greater Bristol Conservation Strategy (NCC 1991). Currently work is being done by Sarah Miles at the Bristol/Avon Regional Environmental Records Centre in collating all records with full documentation of all sites and their status. This work will be of great value to the Planning Authorities in respect of development control and planning applications. The initial aim was to identify geological sites, and as a result, there are only two geomorphological sites included. However, the urban area of Bristol has a varied and exciting geomorphology which lends itself to interpretation and which should be conserved.

The significance of geology and geomorphology in Greater Bristol

Interest in Bristol's geology dates from the early nineteenth century. Buckland & Conybeare (1824) described the geology of the Avon gorge, whilst the Bristol Naturalists' Society, founded in 1862, has a long history of publication dealing with the geology of the region including the seminal paper by Vaughan (1906) on the Carboniferous Limestone Series of the Avon gorge. Geomorphology has featured less, but Harmer (1907) speculated about the origin of the Avon gorge, as more recently, have Bradshaw & Frey (1987).

Geology and geomorphology dominate the urban scene in Bristol for several reasons. Structure has played a significant part in landform development, for example the Kings Weston ridge and the rim of the Westbury-on-Trym anticline form a distinctive watershed around the basin of the River Trym. Much of the high ground within the city is formed of the Carboniferous Limestone which was uplifted during the Variscan Orogeny towards the end of the Carboniferous. Towards the east and northeast of Bristol the Coal Measures dominate the landscape. Most of the low lying areas within the city correspond to Triassic basins. In some areas, the folding and faulting brought the Old Red Sandstone rocks to the surface. Consequently, in a very small geographical distance within the city, representative rocks of four geological systems can be seen, together with their associated structures. Rhaetic and Jurassic rocks overlie the plateau surface in the area of Filton [ST 580 800], whilst to the south Dundry Hill [ST 570 660] has many landslips owing to the weakness of the Jurassic clays underlying the Inferior Oolite. Views from hill tops and high buildings include the Cotswold scarp to the east of the city and the Jurassic outlier of Dundry Hill to the south. The exposure of the older rocks is primarily the result of the development of gorges on the courses of the Rivers Avon, Frome and Trym. Some of the exposures are natural and some the consequence of quarrying. In addition, there are disused railway cuttings and sunken roads which all afford easily accessible exposures.

Geomorphological features

The effects of geomorphological processes are also very evident, because, unlike many cities where the geomorphology is buried under concrete, in Bristol there are many parks and open spaces which coincide with distinctive landforms.

Geomorphological sites in Greater Bristol can be found within both the active and static categories identified in the Nature Conservancy Council's earth science conservation strategy (NCC 1990), though most are static.

Landforms such as periglacial dry valleys, river terraces, active fluvial sites, land slips and the occasional rockfalls as described by Hawkins (1987). The challenge for earth science conservation is that the value and significance of these sites should be appreciated, not only by geomorphologists but by local Planning Authorities and the citizens of Bristol.

Landform databases

Categorization of sites might be according to landform, for example river terrace, hillslope, or according to status, for example good teaching site, rarity, significance in the literature, or value as a classic.

Lists and grid references go part of the way in the compilation of a database, but because landforms are usually areas rather than sites, or because it is the view which is significant, a spatial database is desirable. Wright (1990) describes the documentation of landforms on Ingleborough using a computerized database. It is now possible to extend this approach to surveys of a broader ranging, less detailed type, appropriate for use in urban areas. Digitized contour maps and raster maps of drainage at the scale of 1:10 000 recently produced by the Ordnance Survey will enable landform categorization to be done in a rapid and effective way. Drainage can be overlain on the contour map and by using Autocad, or a similar system, the areas of interest could be rapidly marked on the map. This would provide an easily retrievable database which could be rapidly amended.

It is proposed that a geomorphological map be produced for the whole city using this method. This could then form the basis of a computerized model, and provide the spatial database used for site selection and for planning purposes. Some preliminary work is being undertaken at the University of the West of England, using the Stoke Park, Frome Valley area (Fig. 9.1).

Communication of information

Should such a database be developed the question is raised as to its use. The value is very much dependent upon communication between the interested parties who include planners, teachers, conservationists and academics. An important step in communication was taken at the instigation of English Nature when representatives from English Nature, planners and Conservation Officers from the County and District Councils, members of the regional branch of the Geologists' Association, staff of the Bristol/Avon Regional Environmental Records Centre and some academics met in Bristol with the aim of extending information about RIGS. Such meetings and the sharing of information help to further the cause of earth science environmental education and conservation.

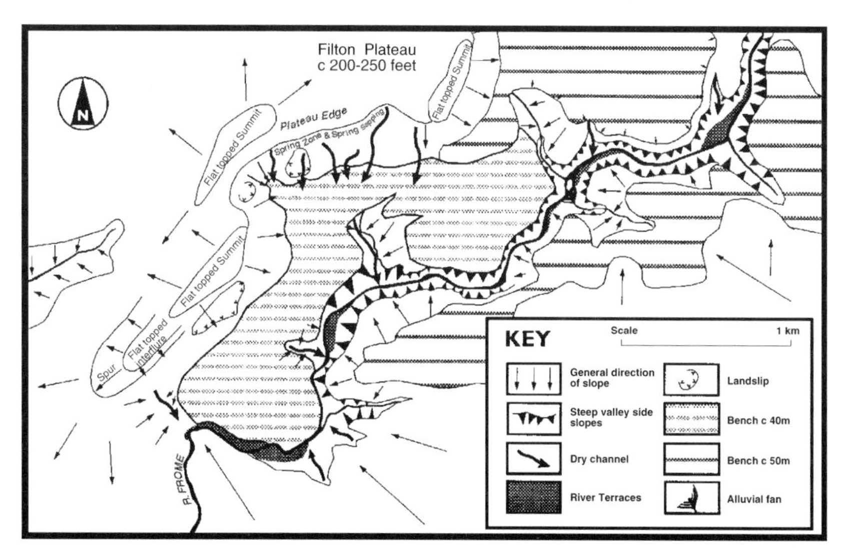

Fig. 9.1. Landforms in the Frome Valley, Stoke Park and Purdown area, Bristol.

Questions asked by the public

Desire on the part of some earth scientists to inform in order to conserve, is insufficient in itself to bring about implementation of either conservation or interpretation. Two important questions are 'what do people want to know?' and 'how can the information be presented in the most exciting way?' In an attempt to answer the first of these questions a survey was carried out during the summer of 1993 at three sites where the landforms are easy to see and some rocks are exposed. Although not in an urban area the sites are similar to sites in Bristol, being the limestone gorges at Burrington and Cheddar and an area of slopes which form a promontory on the coast. Thus the findings have relevance to the gorges, slopes and rock outcrops within the city.

In answer to questions about interpretation 93% said they would read a notice board, 76% said they would follow a trail if it were free, and 64% of the respondents said they would pay for a leaflet.

It would appear that there is plenty of interest in the subject and the problems lie in the lack of provision of information in a user friendly, vandal-proof form. Jarman (1992) suggests several ways of making geomorphology interesting by demonstrating the dynamic nature of the subject rather than merely presenting static information in a leaflet or notice board, though arguably these have their place.

Conservation and education

Strategies to achieve conservation have been published (NCC 1990, 1991) and many innovations have been reported in the journals *Earth Science Conservation* and *Earth Heritage*. Many issues have been considered and should continue to be addressed. These include how to raise public awareness, how to get community involvement, how to work best with the planning authority, how to coordinate and communicate between the various interested parties, how to fund and provide the right kind of interpretative material and how to find out what the public wants with regard to interpretation. The formation of RIGS groups should help to achieve some of the answers to these questions. In Bristol, thanks to the history already outlined, the RIGS sites were identified before the RIGS group was formally established. Action is now underway to establish a group, which it is hoped will provide the focus for communication which is so important.

Within higher education there are opportunities for promoting awareness of landforms and their interpretation. In addition to the usual geomorphology courses in geography degrees, at the University of the West of England we have instituted a group of modules on environmental heritage as options on a part-time Combined Studies degree. Adults with no previous experience in geomorphology are introduced to the study of landforms and their interpretation. By the end of the module students have shown competence in writing a landform trail and an interpretation board. Although they are unlikely to become practising geomorpholgists or professional interpreters of sites, they gain interest and a critical awareness of the landforms around them. In such a course, the local, urban landforms are of particular interest. Interpretation by non-professionals, who nevertheless have some knowledge, is very revealing as the focus is on what has captured the imagination and is perceived to be of interest from a generalist point of view, rather than what an expert thinks ought to be of interest.

Robinson (1990) has exhorted earth scientists to accept a role to add geological understanding to familiar landscapes. Geomorphology should be included as it is primarily owing to the geomorphology that a landscape is distinctive. If these landscapes are in the local park or well known viewpoint within the urban area, the potential for interpretation and education is very high, as large numbers of people visit such a site. Because many are already adapted for public access, the problems of fragility of the site may not exist. The raising of public awareness can be demonstrated by taking a case of local history.

Stoke Park, a case study

Parkland

Within the city boundaries there are several parks, former estates of great houses, which were in a rural landscape before the city engulfed them. They remain as open space where the public can walk and, if so inclined, observe the landforms. Interest in Bristol geology and geomorphology has tended to be associated with the Avon gorge, or on Blaise Castle estate which contains the problematic River Trym, which has a large gorge, but no apparent catchment (Hawkins 1972). There are, however, other landforms in public open spaces which are prominent and equally difficult to explain in terms of their origin.

This case study focuses on one such site and serves to illustrate some of the opportunities which parkland offers for landform conservation. However, the fact that an area is parkland does not in itself afford protection from development.

Stoke Park was originally the grounds of a manor house but in 1740 the land was landscaped as a park and the house remodelled, becoming the dower house for the Beauforts until it was sold to the Burden Institute in 1915. This eventually became part of the National Health scheme and so became Crown property in 1948. Today the land is part of a self standing NHS trust, but the house has been sold.

Geomorphological interest

Geomorphologically this site has several features of interest (Fig. 9.1). A series of valleys drains off the higher plateau, onto a wide bench or terrace of the River Frome. Early paintings of the manor house show that these valleys all contained streams which rose along a spring line at the junction of the White Lias and the underlying clay. The contour crenulations show that these valleys do not drain into the one major dry drainage channel but disappear into a bench, mapped as Mercian Mudstone, which is drained by one, now dry channel, down to the River Frome. There is no evidence of a drainage network on this bench. These channels are dry because of culverting when the park was landscaped in about 1760, but explanation for the bench and the lack of incision by drainage into it, remain open to speculation. One explanation, and the most likely, is that the bench is a sandy facies of the Mercia Mudstone and acts as an aquifer so that the drainage waters seep away instead of forming channels. There is historical evidence that many of the large houses drew their water from this aquifer. This peculiar atrophied drainage system would be completely lost had the area been built upon.

Adjacent to the park is the flat topped interfluve of Purdown. As a landform it has significance comparable to that of Kings Weston Ridge which, unlike Stoke Park, enjoys RIGS status. Most of the interfluves in Bristol have been built upon, but this one is open space, although not guaranteed to remain so. The incorporation of a greater awareness of geomorphology in RIGS schemes is essential if such landforms are to be protected. Jarman (1992) makes the case for landscape protection on the grounds of geomorphology. He suggests that small geological exposures are easily protected by local plans, but big areas such as the one described here, suffer greater threats.

Lessons from history

The survival of Stoke Park and Purdown into the late twentieth century without being built upon is an accident of historical land ownership. In recent years, however, there have been threats to the area.

The first major devastation to this landform assemblage came in the late 1960s with the building of the M32. This cut across the landscaped parkland and brought about the draining of the artificial lake. Further threats came to Purdown during the 1980s when there were plans to sell school playing fields for development which were only prevented by public outcry. Future risks to the area may be demands for more access roads into the massive developments which have taken place on the north Bristol fringe.

Public pressure and publicity helped to save Purdown and Stoke Park from development. It is now within a Conservation Area and the park is a Grade 2 registered historic park or garden. Its value as a landscaped park has been realized and measures are afoot to restore the Duchess's Pond and the monuments. The motivating force for this is the Stoke Park Restoration Trust, a voluntary pressure group.

The issues related to this site demonstrate several principles which earth science conservationists might note. Public awareness of the historic value of the site was aroused initially by one person researching into and publicizing the history of the site. Local pressure group interest was fired when people realized they had something to be valued on their doorstep which was under threat. The ingredients for demands for conservation here appear to be a blend of heritage, nature, culture and amenity. The area is now a Landscape Protection Area, is a site of County Importance for Wildlife, and part contains some woodlands in the English Nature inventory of Ancient Woodland. Perhaps it ought to be a RIGS also, with a description of the geomorphology available to the Restoration Trust and any other interested parties.

Conclusions

This paper has outlined the developments which help to promote earth heritage conservation and more particularly, the potential for geomorphological conservation in Bristol. It is a city with tremendous potential: the landforms are visible, there is a tourist industry, a lively Parks and Leisure Department, an enthusiastic museum service and an Environmental Records Centre. Several problems remain to be addressed with regard to geomophological interpretation and conservation; the time has come to raise public awareness of this aspect of our environmental heritage. They could be addressed by a strategy. First, geomorphologists might become involved with their local RIGS group and Wildlife Trusts in order to raise awareness of geomorphology. Interpretation boards do exist, but the geomorphological interest should be more widely promoted to those designing the display. It is perhaps too late in Bristol to suggest that the impressive stone interpretation plaque for Brunel's Bridge should mention the gorge it spans, or that a notice board on the Downs should mention the geology or geomorphology of one of the most spectacular views in the city. Local history and wildlife conservation only became significant in the planning system through the efforts of enthusiasts; geology is just coming onto the agenda, but has a long way to go; geomorphology lags behind (Pounder & Rose 1992). Heritage and tourism also provide opportunities for the geomorphologists to have their subject valued in the public domain. Geomorphology has an unfortunate name, in that the general public does not meet it very frequently, the origin of many landforms is difficult to explain and often controversial, but the challenge is to geomorphologists to make their subject come alive.

Acknowledgements

Dr Stuart Harding is most sincerely thanked for discussions and information on Stoke Park. Sarah Miles is thanked for providing information on the work of the Bristol/Avon Regional Environmental Records Centre. The Faculty of the Built

Environment, University of West of England is thanked for supporting research and technical facilities for the production of this paper. I am also grateful to Andrew Mathieson for his interest and support.

References

Bradshaw, R. & Frey, A. E. 1987. The geology of the Avon Gorge. *Proceedings of the Bristol Naturalists Society*, **47**, 45-64.

Buckland, W. & Coneybeare, W. D. 1824. Observations on the South-western coal district of England. *Transactions of the Geological Society*, **1**, 210-316.

Harmer, F. W. 1907. On the origin of certain canon like valleys associated with lake-like areas of depression. *Quarterly Journal of the Geological Society of London*, **63**, 470-514.

Hawkins, A. B. 1972. Some gorges in the Bristol district. *Proceedings of the Bristol Naturalists' Society*, **32**, 167-185.

---- 1987. Rock slope stability in the Avon gorge. *Proceedings of the Bristol Naturalists' Society*, **47**, 65-78.

Jarman, D. 1992. Planning Protection. *In:* Stevens, C., Gordon, J. E., Green, C.P. & Macklin, M. G. (eds) *Conserving our landscape.* Conference Proceedings, Crewe, 204-205.

Mathieson, A. 1980. The development of field work resources at Bristol City Museum. *Earth Science Conservation,* **17**, 10-12.

NCC. 1990. *Earth science conservation in Great Britain, a strategy.* Nature Conservancy Council, Peterborough.

---- 1991. *The greater Bristol conservation strategy*. Nature Conservancy Council (English Nature) S W Region, Taunton.

Pounder, E. J. & Rose, J. 1992. Landforms and ice-age inheritance: upper Swaledale, North Yorkshire, UK. *In*: Stevens, C., Gordon, J. E., Green, C. P. & Macklin, M. G. (eds) *Conserving our landscape.* Conference Proceedings, Crewe, 185-190.

Robinson, E. 1990. Seeing with new eyes. *Earth science conservation*, **28**, 24-26.

Vaughan, A. 1906. The Avonian of the Avon gorge. *Proceedings of the Bristol Naturalists' Society*, **1**, 74-168.

Wright, R. 1990. Ingleborough: computerising a landform suite. *Earth Science Conservation*, **28**, 15-17.

10 Museums: a focus for urban geology and geological site conservation

Jonathan G. Larwood & Kevin N. Page

Summary

- Museums can have a key role to play in urban geology and geological site conservation.
- Museums have been the centre of geological collection and study from the early nineteenth century.
- Today, this activity continues, but museums have a key role to play as the link between the local environment and the promotion of urban geology.

Museums have always been a centre for local history and documentation and a repository for locally collected specimens. This has proved equally true for geology as it has for natural history, archaeology and social history. Museums have played a primary role in establishing local geological collections since the early nineteenth century (Knell 1996); first as a centre for the Victorian philosophical society then for the natural history society. Interest in museum geology has seen a fall and rise mirroring the popularity of such societies. Today, the cycle has turned fully and, despite the economic recession, there has been a resurgence of interest in the museum geological collection. This paper examines the current role that museums have to play in urban geology and urban geological conservation; with over 170 geological museums (most urban) museums remain very much a focus for the geology on our doorstep.

The inherited resource

With a long history of involvement in the science of geology, museums often house both historically and scientifically important collections of local, national and international significance. These may include locally obtained palaeontological or mineralogical collections, such as the vertebrate material of Dean Buckland collected from Kirkdale Cave in

Yorkshire in the early nineteenth century (now housed in the Yorkshire Museum), or from further afield, diverse collections such as Wisbech Museum which includes North American fossil insects. Collections may be themed or be broad in scope and often include the spectacular and rare. One of the most appropriate to this volume is the John Watson collection of building stones and materials housed in the Sedgwick Museum in Cambridge (Andrew 1994). Collections have always been a museum's most valuable asset forming a basis for study and education and, through display, the main public attraction. Early displays, however, rarely imparted more than basic systematic, stratigraphical or geographical information which reflected the Victorian desire to collect and to classify in order to begin to understand natural history. It was only a minority of curators, however, who realized the educational use and the potential for wider public entertainment of these collections.

Unfortunately, it is frequently this 'traditional' approach (where it has persisted into the twentieth century) to museum collections and display that has led to the stereotyped view of museums being a collection of dusty and decaying specimens, darkened rooms and cases, looked after by a curator whose lifetime is spent numbering, labelling and cataloguing (Lewis & Foster 1995).

Today, much has been done to redress the balance and to fulfil the needs of a better educated and increasingly enquiring public alongside the maintenance of scientifically and culturally important collections.

The modern museum

Today's museum has much to offer the urban community. Museums are a centre for furthering our scientific knowledge through the provision and conservation of a research resource, a centre for education, at all levels, using this resource and a centre for promoting and even implementing earth heritage conservation.

Science and research

Lewis & Foster (1995) outline one of the primary roles of the museum curator as being 'to ensure that specimens and the information about them are available for examination; to add and update information when necessary; to enhance the collection by the addition of new specimens.'

This clearly defines the key position museums have in facilitating research.

As a repository for geological collections, museums have always been central to research. National and many provincial museums house type collections ensuring specimens, with relevant documentation, are available for present and future study. Equally, many collections are continually added to, some local museums building a reputation for themed geological collections of international importance. At the same time familiarity with collections creates a knowledge of specimens and local geology with which museum staff can often take a lead role in future discovery and research (Radley 1994).

Education

The scientific resource which museums hold supports their educational role at all levels, from general enquiries (from the lay person to the expert) to creating displays, organizing one-off events such as identification days and workshops, organizing and leading field trips, presenting lectures, and producing interpretative literature.

From the Victorian-style display with its desire to classify, to today's display aimed at an increasingly informed audience with a desire for knowledge, the museum display has moved from the traditional to the innovative (Nudds 1995). Knell & Taylor (1989) outline three key approaches to geological display. The geological approach, an accessible summary of local geological history; the biological approach, a comparison between fossil and modern fauna and reconstruction of past environments; and the historical approach, outlining the history of research into local geology and documenting the history of exploitation of local mineral resources.

Success seems to come when these elements are combined. The 'Time Trail' in Dudley Museum and Art Gallery brings together the geological history of the Dudley area from the earliest Silurian seas through to the last ice age. Not only are past environments represented but the three dimensional displays show how fossils and rocks are formed and are linked to today's industrial exploitation of this resource. The National Museum of Wales in Cardiff (with greater financial resource) has developed the 'Evolution of Wales' exhibition that incorporates sight, sound and smell with erupting volcanoes and robotic dinosaurs.

Nevertheless, the more 'traditional' specimen based approach is still very important for visitor reference and geological teaching, for instance, the chronologically arranged collection of the Sedgwick Museum in Cambridge. At the Hunterian Museum in Glasgow a recent addition of dinosaur eggs to the collection attracted many visitors to see the curators gradually revealing the eggs, perhaps the ultimate in living displays. Here, parallels can be drawn with the Archaeological Resource Centre in York where every display is interactive, visitors can watch research in action and can actually sort samples, adding to the research effort. Like geology, archaeology is about discovery (finding things, close observation, discrimination and detection). Interactive centres can create an understanding and appreciation of geology, good examples being the Charmouth Heritage Coast Centre on the Dorset coast and the National Stone Centre in Derbyshire.

Some museums have 'school rooms' specifically designed to act as classrooms in which museum staff can talk to or teach groups of all ages on specific historical, cultural or scientific subjects including geology. Many of these are occasional events, but others can be part of planned educational programs. Use by interest groups, including supportive 'Friends' of the museum organizations, is also common. Some museums, often in association with area museum services, also produce packaged displays which either travel between museums or can be taken directly into schools, particularly important with today's inclusion of the earth science within the National Curriculum (Hawley 1996).

Events

These can be important for promoting the relevance and understanding of local geology as well as the understanding of geology as a subject. One of the most successful events in recent years has been the Dudley Rock and Fossil Fair, organized by Dudley Museum and Art Gallery, first held in 1992 and again in 1994. This event has provided the opportunity for local and national geological societies, universities and schools, conservation organizations, industry and collectors to set up stalls and displays aimed at widening (whether directly or indirectly) a public understanding of geology.

Equally, such events can also exploit current issues, note 'Jurassic Week' held at Peterborough Museum and Art Gallery in 1993, its establishment leading on from the popularity of Steven Spielberg's film *Jurassic Park*. Events during the week included fossil identification, guided geological walks, lectures, films and talks, and children's workshops constructing Jurassic inspired puppets with the locally based foundation Puppet Works.

Exhibitions (including travelling displays) are the main stay of education in most museums while events aim to attract, motivate and involve a wider range of visitor. Both demonstrate the central role a museum has in furthering our understanding and support for geology, within and beyond the urban environment .

Museums and site conservation

Clearly recognized as being central to research and education, museums have an equally important role in earth heritage site conservation. Conservation of the collection 'is to ensure the continued existence of the specimens and data concerning them' (Lewis & Foster 1995). Lewis & Foster conclude that research, curation and conservation (specimen) should be regarded as an integrated activity, the common goal being the furtherance of our scientific knowledge.

Museums have always been involved in conservation beyond specimen conservation. The relationship between museums and collectors has always been strong. Early collectors furnished museum collections, but importantly this association developed the techniques which are so important for successful excavation and documentation of relevant information (Knell 1994).

Today the relationship between museum and collector is still maintained. For example, at Peterborough Museum and Art Gallery, amateur geologists, together with museum staff, have spent many hours excavating, preparing and putting on display spectacular marine reptiles from the Peterborough Jurassic Oxford Clay brickpits (Chancellor 1994). At Conesby Quarry, Humberside, Scunthorpe Museum has developed an agreement with collectors (with the consent of the quarry operators and land owner) working the fossil-rich Liassic Frodingham Ironstone. A 'shopping list' has been agreed ensuring that specimens absent from the museum collections are donated to the museum. During the period that the scheme has been in operation, many specimens have been added to the collection, perhaps most notably, the first articulated crinoid from the Frodingham Ironstone.

As well as forming close links with both collector and quarry operator, museums have an important advisory role in order to ensure that specimens are responsibly collected and curated. There are general guidelines such as the 'Thumbs-up' guide 'Rocks, fossils and minerals - how to make the best of your collection' published by the Geological Curators' Group, or more specific guidelines such as 'Guidelines for collecting fossils on the Isle of Wight' available from the Museum of Isle of Wight Geology, which is aimed at giving tailored advice to any visitor to the Isle of Wight intending to collect fossils. Equally, direct involvement of museums in specimen excavation continues to be crucial; for example, the recent recovery of a Lower Cretaceous sauropod specimen from the Isle of Wight would not have happened without the vigilance, expertise and man power of the Isle of Wight Museum of Geology (Radley & Hutt 1993). More recent specimen rescue has included the Yorkshire Museum excavation of a Liassic crocodile from the North Yorkshire Coast and the on-going excavation of the Pleistocene 'West Runton Elephant' by the Castle Museum in Norwich.

Many museums also act as Geological Recording Centres which were established as part of the National Scheme for Geological Site Documentation (NSGSD). Such museums are uniquely placed, through links with the local authority, to monitor potential development threats and, where appropriate, to provide guidance on geological conservation. Many, such as that at Bedford Museum, are closely linked to the specimen resource. Here indeed, the establishment of the records centre, including site survey work, added a significant amount of new material to historically important collections.

The Geological Recording Centre set up in Dudley in 1988 (Reid 1994) was based on site recording by the Black Country Geological Society. In total, 340 records are entered including one National Nature Reserve (Wren's Nest), six SSSI and 25 Sites of Importance for Nature Conservation (Box & Cutler 1988). The museum's co-ordination of the Geological Recording Centre has been vital in establishing closer links between the local authority's Planning and Public Works Department, the museum and the Black Country Geological Society. The additional development of a Geographical Information System which uses these data makes the comparison of planning information and conservation information more readily accessible. Developments impinging upon known sites (statutory and non-statutory) can be flagged up at an early stage and appropriate advice given.

What has in the past been informal cooperation between local authority and museum is now being formalized by Dudley Museum and Art Gallery (Reid 1996) into a Code of Practice. This will ensure that geological conservation is a consideration in any development likely to effect or create a new geological exposure, whether temporary or permanent.

Museums are therefore a focus for local geological conservation, through advice to individuals or groups carrying out fieldwork and through documentation of sites (often providing the basis for local site designation) and through links with the local authority. This involvement can be vital for guiding local policy decisions which favour geological conservation.

Conclusion

Museums form a clear focus for urban geology and geological conservation. They provide a valuable resource for the scientific community and are central to geological education. Indeed, for the beginner geologist or local person museums will always act as a focus for enquiry, being a clearly identifiable source of relevant expertise. They have links with and can draw links between many local groups, perhaps most importantly, the local authority.

This central role provides museums with a wide sphere of influence in terms of earth heritage conservation. As well as establishing an understanding of geology, and its value and link to the wider community, museums can provide advice to site users and, in particular, local developers who may affect geological sites. Most importantly, as in the case of Dudley, museums can be influential in local authority policy formulation - the key to ensuring the future of urban geology and geological conservation.

References

Andrew, K. J. 1994. John Watson and the Cambridge building stone collection. *Geological Curator*, **5**, 303-310.

Box, J. & Cutler, A. 1988. Geological conservation in the West Midlands. *Earth Science Conservation*, **25**, 29-35.

Chancellor, G. R. 1994. Pliosaurs and volunteers. *Geological Curator*, **6**, 75-81.

Hawley, D. 1996. Urban geology and the National Curriculum. *This volume*.

Knell, S. J. 1994. Palaeontological excavation: historical perspectives. *Geological Curator*, **6**, 57-69.

---- 1996. Museums: a timeless urban resource for the geologist? *This volume*.

---- & Taylor, T. S. 1989. *Geology and local museums: making the most of your geological collection*. HMSO, London.

Lewis, D. N. & Foster, T. S. 1995. Curation and conservation - the poor relations of research. *Geological Curator*, **6**, 129-132.

Nudds, J. 1995. Geology and the glass case. *Bulletin for the Centre for Environmental Interpretation*, **8-9**.

Radley, J. D. 1994. Collecting dinosaurs on the Isle of Wight, Southern England. *Geological Curator*, **6**, 89-96.

---- & Hutt, S. 1993. The Isle of Wight sauropod. *Earth Science Conservation*, **33**, 10-12.

Reid, C. 1994. Conservation, communication and the GIS: an urban case study. *In*: O'Halloran, D., Green, C., Harley, M., Stanley, M. & Knill, J. (eds) *Geological and landscape conservation*. Geological Society, London, 365-371.

---- 1996. A code of practice for geology and development in the urban environment: a new Local Authority initiative. *This volume*.

11 Museums: a timeless urban resource for the geologist?

Simon Knell

Summary

- The historical development of museums and their collections is reviewed.
- Museums played an important role in the early popularization of geology.
- It can be demonstrated that museums and their collections follow a cycle of growth and decline as their role within both the local and scientific community changes.

One of the most important geological resources on our urban doorstep is the local museum and its geological collection. Museums exist in almost every urban landscape in Britain. In a high-tech age museums often present an image of being outmoded and no longer contemporary. In the last few years, however, geology in provincial museums has undergone something of a renaissance and many have found sufficient funds to renovate their tired displays. Those museums which possess a geologist remain at the heart of amateur activity, a role many first took on over a century ago. Museums provide urban resource centres and draw on a wider geological community committed to communicating geology to the general public and supporting the progress of British geology. They are therefore an important part of our urban geological resource.

These urban museums evolved, however, in another era, in response to very different needs and opportunities from those that exist today. Many date from the very earliest days of geology, its 'Heroic Age' in the nineteenth century. Since this 'Heroic Age' geology has evolved dramaticaly but how has the geology in the museum evolved with it and how has the role of the museum within the local community changed? The aim of this paper is to consider these questions and to chart the history of geology within urban museums.

Museum geology in the Heroic Age

In the early nineteenth century geology began to develop its own framework for scientific study. As the century progressed it became an increasingly powerful magnet for the cultured and wealthy classes, and for the scientific elite. It revealed a British landscape extraordinarily rich and diverse in geology, but largely unexplored. For the embryonic scientist/philosopher the opportunities for discovery were boundless, and the products of these discoveries remarkable. By the 1820s geology was starting to become extremely fashionable, a fashion which continued to develop over succeessive decades. Among the country's upper echelons an interest in, and knowledge of, this science became an essential sign of sophistication.

In part, this popularity was fed by the remarkable discoveries of these decades, of marine and flying reptiles, of dinosaurs, of hyenas once living in Yorkshire. For the informed local gentry the regional works of Gideon Mantell, John Phillips and others were beginning to provide a framework for understanding local fossils and rock successions. Palaeontological and geological syntheses, including elementary texts, presented a new and developing science to which anyone could contribute and a science that courted both controversy and lively public debate.

Geology, unlike the inaccessible and abstract sciences of mathematics, astronomy, physics and chemistry, was tangible, comprehensible, romantic and everywhere. It had yet to accrue its plethora of jargon terms. Its ideas, which were the products of simple inductivism, were conveyed in elegant descriptive prose often tinged with varying degrees of Romanticism. Geology permeated deeply into the leisure activity of the cultured. Morrell (1994) suggests that the 'perpetual excitement' surrounding geology was in part due to its economic benefits, its interest to the traveller, its adaptation to any scale of study and to its relationship with religion.

It was in this scientific climate that the many existing museum collections were formed. Geological advance and opportunity, combined with widespread interest in natural history, gave birth to the literary and philosophical movement which swept through much of Britain, but which was particularly strong in Yorkshire. Provincial Britain offered such unparalleled opportunity for geological research that every rock in every parish had the potential for turning this new science literally upside down.

The new provincial societies could have followed a number of national models but more than any they emulated the Geological Society which, since its establishment in 1807 had become not only the liveliest debating house in London (Rudwick 1985) but also the national repository for collections associated with geological advance. For a few decades a geological research network was to spread throughout Britain. Established in expanding urban centres, the provincial learned societies provided a focus for the intellectual pursuits of the growing middle classes, the local gentry, medical men and clergy. They played a vital role in feeding local intelligence to higher science, and in turn were rewarded with up-to-date intelligence, collections or simply association. The new and fashionable science of geology created the excitement that could keep a society enthralled. The members of the societies, eager to establish their own immortality, built collections which fed the research machine. The interaction was both symbiotic and catalytic. Never have museums been so closely intertwined with pioneering contemporary science.

While the societies conformed to a general pattern and shared similar objectives and methods (Allen 1976), they also had their differences. These were primarily a reflection of local personalities, and the size and composition of the social strata from which they were formed. Relative status could be measured in the size of the museum building, the number of members or its links with the scientific elite.

In a county with a society in every urban centre of any size, the Yorkshire Philosophical Society reigned supreme. The societies of the smaller towns of Whitby and Scarborough were definitely second tier despite having on their doorstep some of the most accessible fossil resources in Britain. There were simply insufficient numbers of local philosophers in these small towns to provide the critical mass necessary for success on the scale of that seen in York. The York society also benefited from dynamic and influential members including William Venables, Vernon Harcourt and John Phillips who, in particular, rose rapidly from humble beginnings to become a central figure in British geology. York was a centre for science, and while valuable collections and expertise developed on the coast, these smaller and poorer societies often relied upon the York philosophers for intelligence and contacts with mainstream science.

For the Yorkshire Society there was no more important objective than revealing the geology of the county. To achieve this it reached out from its urban base to gather geological specimens from all parts of Yorkshire. Increasingly powerful and assertive in its influence, the Yorkshire Museum became a particularly attractive repository for the finds of collectors in preference to the smaller, more local, society collectors. Placed in the York collections, there was a real chance that the specimen would be seen and described by the science and reflect glory on the finder.

By 1850, however, the heyday of the philosophical society was coming to an end. The science of geology was becoming increasingly rigorous and systematic, and the publications produced were less approachable and more specialized. At the heart of this professionalization was Henry de la Beche's Geological Survey which was establishing a new level of resolution in data collection. The philosophical societies and the museums they created began to founder. The provincial youth were not attracted to these ageing and increasingly conservative gatherings, and indeed society itself had moved on.

Scarborough Literary and Philosophical Society provides a particularly dramatic example of the life-cycle of these institutions. The gentry of Scarborough had been seriously discussing the establishment of a museum

Fig. 11.1. The Rutunda museum, Scarborough. Built to house the collections of Scarborough's Literary and Philosphical Society (Photograph: P. Doyle).

in the town since 1820. Within two years of the establishment of the Literary and Philosophical Society in 1827 the remarkable Rotunda was built (Fig. 11.1), a building constructed in the round so as to provide the best opportunities for the stratigraphical arrangement of fossils. However, the building, which was soon considered too small for meetings or the display of the growing collections, placed a heavy burden on the Society funds. Constantly short of money, the society waited nearly a decade before the specimen cases could be fitted, and when important and desirable local fossils became available they were unable to purchase them. By 1842 the Society was approaching crisis and desperately attempted to attract new members. When in 1848 the loan on the building was recalled, the embarrassed state of the institution became publicly revealed. The museum was described as being disorganized, poorly labelled and unattractive. Membership had been declining since 1836. Less than a decade after the institution was founded it was close to collapse. The Society only managed to survive by merging with the town's archaeologists in 1853.

These declining institutions became the brunt of considerable criticism throughout the latter half of the nineteenth century. They lacked organization, control of collecting and informed interpretation (Manton 1900; Scharff 1912). Some philosophical societies survived, but their museums became considerable burdens and most were eventually passed to other bodies.

The second coming

It is far easier to establish an interest group than to maintain it. What is new is often fashionable, there is glory in the founding and considerable control in defining its objectives. After a time fashion and glory are lost, and growing conservatism prove a major obstacle to change. This was the case when interest in natural history revived in the 1860s. Rather than modernize the philosophical societies, new natural history societies and field clubs were established (Anon 1870; Allen 1976). While many of these wished to take advantage of new travel opportunities, and did not at first wish to be encumbered with museums, these bodies soon found that museums were a vital component in assisting in the study of natural history (Anon 1871). Unlike their philosophizing predecessors, however, these new societies were more likely to pressure their local authority into establishing a museum. This became a function of natural history societies even into the twentieth century. Many of these societies were established in towns which had been too small to participate in the philosophical movement forty years earlier. As the century progressed and the urban centres grew, a critical mass was achieved which burst forth as a natural history society. Geology - its materials so easily collected and preserved - became a key element in their activities and the museums they encouraged.

The development of these new museums and their relationship to the new natural history societies are well illustrated in Canterbury (Anon 1871). Canterbury Museum was established by the town in 1847, one of the first under the Museums Act of 1845. Its rapid establishment had been achieved by taking over the town's moribund philosophical society collection, an approach already adopted by Folkestone Natural History Society. It was this that spurred the secretary of the local East Kent Natural History Society, George Gulliver (1871) to attack provincial museums in general. 'The majority of them throughout England present such examples of helpless misdirection and incapacity as could not be paralleled elsewhere in Europe.' The new breed of society was intent on following an entirely different tack to those which originally founded the collections. Now the buzzword was education and the museum they sought to create for this purpose had no use for the type of collection popular with the philosophers.

The material in Canterbury and other similar towns which had survived the decline of the local philosophical society, and the neglect of the town council, was now to be streamlined by yet another group. In rescuing the useful material a good deal of 'rubbish', in the eyes of the new group, was likely to be encountered. According to Gulliver (1871) the solution was simple 'sell it if you can, or give it away; but by all means get rid of it, and that swiftly; to which end a bonfire might be the best thing.' While local collections might survive, important material brought back from further afield was often considered junk.

After this boom of interest in geology and natural history around the turn of the century the science entered another period of decline (Boswell 1941). The loss of material from the 1920s onwards, some dating back to the earliest days of local geological exploration, was considerable. At this time there was no safety net: museums had not been sufficiently professionalized to be concerned about disposal and curators were often uninformed about geology. In southeastern England, Raymond Casey, of the then Institute of Geological Sciences, was a witness to the destruction and rescued considerable amounts of material including virtually the whole of the collection from Tunbridge Wells (Gill & Knell 1988). The neglect of geology collections, and in some case their loss, was widespread. In fact Doughty (1979) suggested that 'one of the bleakest periods' in the history of the United Kingdom's geological collections occurred during the early twentieth century.

At this time numerous writers were attempting to explain and reverse this decline (North *et al.* 1941; Knell & Taylor 1989). North (1942) saw the decline of geology as a result of a failure to understand and interpret geological material. It was difficult to communicate with geological material in the same way that it was possible to convey ideas with items of greater aesthetic or popular appeal. Allan (1942, p. 58) suggested that the inherent durability of geological materials worked against the interests of these collections because they required 'less attention, they got it, and, while the passage of time saw new material and new methods of preparation and exhibition introduced into other natural history departments, the rocks and the fossils remained intact and inert, sometimes almost invisible beneath the gently accumulating layers of dust'.

Geology and the local museum

The collections of the natural history movement became the inheritance of many provincial museums. The role of the museums was changing at the close of the nineteenth century. The local museum was in many cases being redefined as the educational museum. The museum founded by Hutchinson in Haslemere provides an example. Rather than gather local resources from its hinterland it aimed to diffuse knowledge, an ambition which is embodied in the following statement: 'The object of an educational museum should be to educate rather than collect' (Hutchinson 1893). Flower (1893 *in* Flower 1898) also recommended that municipal, school and village museums should only collect material of general educational value.

Many museums were attempting to reach out to those in the community who, through education and opportunity, lacked the chance of profound study but might be more generally engaged by museum objects (Flower 1889 in Flower 1898). Old snobberies persisted, however. Woodward (1900), for example , felt 'that the "man in the street" did not at present seem to be a very hopeful subject in London. He came into museums chiefly for warmth and shelter, and usually brought a good deal of dirt in with him' (Manton 1900).

Despite being a rebellion against the research orientated museums of old (Hooker 1869), the new educationally orientated geologists felt themselves unable to release their exhibits from the constraints of formal science. A major preoccupation of the early meetings of the Museums Association was how geological collections should be laid out in museums. In the decades when geology was fashionable and fairly exclusive there was no apparent need to communicate effectively; museum displays illustrated current items of news and a curator or member was always on hand to inform the visitor. The philosophical museums attempted little more than to provide the vocabulary of the contemporary science (Yorkshire Philosophical Society 1828). Now, however, even though the museum-going public was largley uninformed about science, the exhibits were more often than not conservative and inflexible. Arrangements were generally systematic, geographical or stratigraphical (Rudler 1877; Dawkins 1890, 1892).

Hutchinson (1893) was one of the few curators who began to break the traditional mould by promoting the liberal use of pictorial illustration, models, casts, descriptive labels and the availability of reference texts in educational museums.

Smith (1897) expressed the benefits of organizing a museum display along the logical organization of a textbook to meet the needs of local schools. A point which was spelt out more precisely by Bather (1924) who had earlier suggested that museums should not be given over to rows of specimens but to ideas, such as Darwinian evolution, which made people think (Bather 1896). Bather's ideas were further developed by North (1928) who suggested that the failure of geology to communicate was due to its lack of connection with everyday life or common knowledge. Equally, there was no longer a sense of wonderment and an apparent absence of utility.

Systematic displays were still prevalent in museums, and were described by North (1942) as having 'a system but no soul; the specimens have names but are devoid of meaning; they demonstrate facts, but they do not tell a story.' It was this importance of storyline which North stressed. Like the supporters of educational museums, and Flower and Bather before him, North saw the fundamental role of museums as the teaching of geological principles. For this, a general collection remained a requirement, local material was an appendix to the general story and to be only used to give the story some local flavour.

North (1928) put forward the idea that a general introductory case describing the main features of geology was essential and should be supported by material which illustrates and explains the fundamental principles. Even so for some, and for Bather in particular, this interpretation was seen as rather dry and technical (North 1931). North's approach was, however, a marked improvement on anything that had been seen before and it found its way into many British museums where it dominated display until the 1970s (North *et al.* 1941).

The overpowering didacticism of the educational movement in museums at least led to thematic and story telling displays. It believed, however, that displays should be purely educational and should teach general geology. As a consequence the same displays were seen everywhere, for example, the stratigraphical column arranged in a series of compartments, geological time as a clock, the biological classification of fossil animals, and the characteristics of minerals.

In the late twentieth century the provincial museum needed to rediscover its own identity: local geology was for many museums the reason why they existed. To rediscover the local environment was to rediscover the museum's roots in earlier philosphical societies where local geology was at the forefront. This rediscovery of local geology has done much to revitalize the collections and displays of our local museums, reversing the decline of the first part of the century.

Museum geology in the future

Museum geology has ridden a rollercoaster of popularity and decline (Knell 1995). The geological collections of urban museums have undergone periods of growth and decline. In many cases neglect has immediately followed, or even accompanied, the growth of a geological collection. The philosphical societies of the first half of the nineteenth century generated collections which should lie at the heart of the science but rapidly fell into decline. These collections were subsequently purged and rebuilt along different lines by the natural history societies, but these too were also liable to rapid decline (Flower 1898; Torrens & Taylor 1990). It is the remains of these collections that form the resource available in many of today's museums and once more they are the subject of revitalization.

Recent years have seen a resurgence of geology within provincial museums. This success is, however, all too dependent on the presence of a geologist within the museum. History shows that museums will not increase geological provision if there is not local demand; but it also shows that it is hard to generate local demand if there is no provision (i.e. no museum geologist). This is the *Catch-22* which faces museum geology. The solution is simple: as geologists we must loby for a geologist in every museum and support the geological work of our local museum. In this way we can expand and retain this valuable urban geological resource.

References

Anon. 1870. Natural history in schools. *Nature*, **2**, 249-250.

---- 1871. Natural history museums. *Nature*, **3**, 381-382.

Allan, D. A. 1942. President's address for 1941: geology in museums. *Proceedings of the Liverpool Geological Society*, **18**, 57-69.

Allen, D. E. 1976. *The Naturalist in Britain: A Social History*. Penguin, Middlesex.

Bather, F. A. 1896. How may museums best retard the advancement of science. *Report of the Proceedings of the Seventh Annual General Meeting of the Museums Association,* 92-105.

----1924. Fossils as museum exhibits. *Museums Journal*, **24**, 132-141.

Boswell, P. G. H. 1941. Anniversary address of the president: part 1: The status of geology: a review of present conditions. *Quarterly Journal of the Geological Society*, **97**, XXXVI-LV.

Dawkins, W. B. 1890. On museum organisation and arrangement. *Report of the Proceedings of the First Annual General Meeting of the Museums Association*, 38-45.

---- 1892. The museum question. *Report of the Proceedings of the Third Annual General Meeting of the Museums Association,* 13-24.

Doughty, P. S. 1979. The state and status of geology in United Kingdom museums. *In:* Bassett, M. G. (ed.) *Curation of Palaeontological Collections*. Palaeontological Association, Special Papers in Palaeontology, London, 16-26.

Flower, W. H. 1898. *Essays on Museums*. Macmillan, London.

Gill, M. A. V. & Knell, S. J. 1988. Tunbridge Wells Museum: geology and George Abbott 1844-1925. *Geological Curator*, **5**, 3-16.

Gulliver, G. 1871. On the objects and management of provincial museums. *Nature*, **5**, 35-36.

Hooker, J. 1869. Address of the president. *Report of the British Association for the Advancement of Science for 1868*, lviii-lxxv.

Hutchinson, J. 1893. On educational museums. *Report of the Proceedings of the Fourth Annual General Meeting of the Museums Association*, 49-63.

Knell, S. J. 1995. The rollercoaster of museum geology. *In*: Pearce, S. M. (ed.) *Museums Exploring Science*. New Research in Museum Studies, Athlone, London, in press.

---- & Taylor, M. A. 1989. *Geology and the Local Museum*. HMSO, London.

Manton, J. A. 1900. A rambling dissertation on museums by a museum rambler. *Report of the Proceedings of the Eleventh Annual General Meeting of the Museums Association*, 65-80.

Morrell, J, 1994. Perpetual excitement: the Heroic Age of British geology. *Geological Curator*, **5**, 311-317.

North, F. J. 1928. Geology and the museum visitor. *Museums Journal,* **27**, 271-276.

--- 1931. Geology in relation to the small museum. *Museums Journal*, **31**, 8-17.

--- 1942. Why geology? *Museums Journal*, **41**, 249-256.

----, Davidson, C. F. & Swinton, W. E. 1941. *Geology in the Museum*. Museums Association/Oxford University Press.

Rudler, F. W. 1877. On natural history museums. *Y Cymmrodor*, **1**, 17-36.

Rudwick, M. J. S. 1985. *The Great Devonian Controversy*. University of Chicago.

Smith, H. J. 1897. Popular museum exhibits. *Report of the Proceedings of the Eighth Annual General Meeting of the Museums Association*, 63-68.

Scharff, R. F. 1912. *The aims and scope of a provincial museum*. Belfast Public Art Gallery and Museum, Publication, **31**.

Torrens, H. S. & Taylor, M. A. 1990. Geological collectors and museums in Cheltenham 1810-1988: a case history and its lessons. *Geological Curator*, **5**, 175-213.

Woodward, H. 1900. Presidential address. *Report of the Proceedings of the Eleventh Annual General Meeting of the Museums Association*, 25-44.

Yorshire Philosophical Society. 1828. *Annual Report of the Yorkshire Philosophical Society for 1827*. 14.

12 Digging up your doorstep: engineers and their excavations

Graham J. Worton

Summary

- Engineering projects in urban areas hold much potential for urban geology through the creation of both temporary and permanent geological exposures.
- The opportunity for geological gain within these projects is immense, but is frequently not exploited because local geologists usually become involved in the planning process only at an advanced stage.
- The key to success is early involvement: geologists need to take a more proactive role in the planning process to maximize the potential for geological gain within urban development projects.

How does a green field become a busy street? Why does one building get demolished to make way for yet another? What opportunities are created for geology when these things happen?

At its simplest development happens because someone in a position of wealth or influence desires it or it occurs in response to social pressures. We all can have a say in what happens in our local area and have the right to object to development projects. Development is controlled through a framework of planning legislation and regulations. These are interpreted and applied by the local authorities in response to local needs and circumstances. Conditions relating to particular circumstances or locally significant features can be attached to planning permissions for development on a site specific basis. This is where geology and geological conservation can become part of the planning process.

A chain of opportunities

Opportunities exist to attach conditions that relate to geology to the planning permissions for new developments. To do this, however, the local importance of geology and geological conservation must be understood clearly by the Planning and Development Control Officers of the Local Authority. Sadly, these officials rarely appreciate the importance of geology and therefore opportunities are frequently lost.

After an idea or scheme has received planning permission, development usually proceeds along the following route: site investigation; design of project; submission for further scrutiny (e.g. building regulations, statutory undertakers etc.); detailed design and specification; tender process at which point all quantities are approximated and quotations sought from companies to carry out the works; contract is let and engineer and contractor are appointed; site work begins.

Opportunities exist at all stages of this process to ensure that a geological component is built into the scheme. In the majority of cases, however, it is only at the site works stage that geologists get involved. At this point they have least influence and there is also the least amount of flexibility in the development process. Geologists are forced into retrospective action often salvaging the bits of geology which are left. It does not have to be this way if, as geologists, we can enter the development process earlier.

The engineer: a curious beast

If site works are where geologists are currently most likely to get involved then it is important to understand the potential role geologists have to play. This is particularly true when tackling the on-site engineer. Except in rare developments which impinge on the margins of statutory sites, such as Sites of Special Scientific Importance (SSSI) or National Nature Reserves, or where geological planning conditions have been imposed, the on-site engineer has no obligation to the geology or geological features which might be exposed during the development work. As a consequence any geological gain (access to record, collect or educate) will be dependent on the goodwill of the client and the engineer.

Engineers and geologists have important differences in philosophy and activity. The engineer is charged with the responsibility of achieving a defined end-product: for example, to construct a building, road or landfill. He/she is bound by a contract, a code of conduct and a timescale. A site engineer's success depends on his/her ability to make things happen to a specific design on a fixed and rigid timescale. Geologists on the other hand, thrive on uncertainty. They never really know what will be found or what the end-product will be. The success of geologists often depends on flexibility, creativity and imagination. From the engineers point of view geologists have the potential to interfere with the smooth running of a project and are therefore tolerated but not truly welcome.

Opportunities for geology are therefore often controlled by the on-site engineer (Worton 1994). Consquently the geological gain depends on what the engineer knows of geology and how we as geologists conduct ourselves while on the site. Geologists must provide the site-engineer with a clear definition of what is required and must respect his/her responsibilities if the potential opportunities are to be maximized.

Geology and engineers: some opportunities

The full spectrum of opportunity that exists from engineering work relates to the primary functions of engineering in the urban environment. These are summarized below.

- Existing structures: repair, improvement extension, including services and structures.
- Site investigation: exploration and assessment for design purposes.
- Exploitation of mineral wealth: location, assay, extraction.
- Environmental protection: waste disposal, water resources, reclamation, defence against natural disaster.
- Earthworks and new construction: ground preparation, tunnels, roads, structures.

Each of these facets of urban engineering has many possibilities for geological endeavour.

Fig. 12.1. Nature in the brickwork. Artificial structures and repairs can mimic geological features. Here at Canal Bridge in Netherton, West Midlands the brickwork resembles an angular unconformity.

Existing structures and municipal works

This includes many existing engineering structures including footpaths, buildings, kerbstones, shop fronts, waterways and gravestones. If we look a little more closely at these urban structures they may have other inherent geological value (Fig. 12.1). These are, and will continue to be, the most significant part of the urban geological resource. They are well established, well known and valued by the urban community.

Site investigation: breaking new ground

The largest financial risk in construction comes from a lack of knowledge about the properties and behaviour of the ground. It is not surprising therefore that the urban geological community is frequently involved in providing site assessment information for development sites in urban areas. In practical terms, this means drilling boreholes and digging pits and trenches to examine, sample and test the ground. Enquiry of this sort is specific to the needs of the client, is competitive and has a very rapid turn around time from hole-to-report. It is no surprise then that aspects of 'geology' such as sedimentology, palaeontology and mineralogy are rarely considered in engineering geology. Opportunities exist for the wider geological community to exploit this resource. Site documentation, cores (Fig. 12.2) and excavations may be available for study given the goodwill of the developer and the promise of confidentiality. Samples will inevitably outlive their shelf-life as there is a continual influx of new material to the engineer's storerooms. Engineers, in the author's experience, would be happy to see non-confidential samples earmarked for disposal go to a good home such as a school, university or museum. The British Geological Survey has made very good use of this resource.

Location, assessment and extraction of minerals

Again, site investigation is also usually a necessary first step in the exploitation of mineral reserves. Confidentiality is, however, much more likely to be a barrier. Similar rules and goodwill apply to accessing minerals excavations, but by their very nature, such excavations are often designed to have a pay-back over several years or even decades. This means that opportunities to view exposure are greater and each return

Fig. 12.2. Engineers' core and samples. The photograph shows cores of Silurian limestone from Castle Hill, Dudley.

visits may expand the knowledge of a site. Every urban area will have a number of quarries in or around it providing a very valuable earth science teaching resource.

Environmental protection

Increasingly engineers are involved in a wide range of environmental protection works. These include a variety of earthworks and construction such as landfill, flood defences, noise barriers, cleansing of contaminated land and ground stability works.

Many of these projects require the bulk movement of soil and soft rock. These materials are commonly spread or mounded at the surface. They may, as recently seen on the Dibdale/Burton Road land reclamation project in Dudley, be highly fossiliferous shales. These were spread for landscaping purposes and have weathered to yield numerous Silurian fossils. They may also provide superb petrological laboratories when glacial till is used in landscaping with its range of exotic erratics (Fig. 12.3). Opportunities may exist anywhere that stone is being imported or soils moved for landscaping.

Fig. 12.3. Many soft rocks are used for environmental engineering earthworks. The photography shows glacial tills being used as a landfill lining. The till contains exotic erratics from Scotland, northern England and Wales.

Construction

New construction inevitably interacts with the ground. The scale of excavations may vary considerably from small trenches for foundations or services to huge earth moving projects. Exposures created may be temporary or permanent. The scope here for creative design and planning to maximize the geological potential of a site is rarely exploited.

Smaller excavations may never penetrate deep enough to encounter natural ground but merely expose and terminate within ash, foundry sand and other industrial discards collectively known to the engineer as 'made ground' or 'fill'. Even in these materials a geological story may be told about use of minerals. Most large urban centres with long industrial histories will have widespread deposits of such materials covering the solid geology below. Small excavations are numerous within urban areas and many do not require planning permission. Keeping informed of them, gaining access to them and having time to log and collect is a near-impossible task, and certainly not an exciting or glamorous one.

Large urban engineering projects on the other hand tend to be deeper and last longer. These offer many more possibilities for return visits for recording, collecting and education. It is often possible to use such projects to develop good engineer/geologist relationships for the future.

Black Country case studies

Case study 1

The Dibdale/Burton Road Land Reclamation Scheme in Dudley illustrates what can be achieved in large engineering projects through fostering good relationships. The site, 1 km to the northwest of Dudley town centre, occupied 27 hectares and had widespread and complex problems of contamination, waste disposal and instability due to mine shafts and mine workings which used to exploit coal and fireclay seams. The reclamation scheme involved open cast coal mining to extract the remaining coal, removal of the mine shafts and the creation of a void. The void was lined using the available clays and mudstones to provide a repository for the waste and contaminated materials on the site. The works created large exposures of Coal Measures rocks and exposed spectacular room and pillar mine workings. Liaison with site staff allowed small parties to visit but the major geological gains were made by the professional engineering staff themselves.

The project produced a very detailed geological log of lithology, sedimentology, stratigraphy and palaeontology. Several new fossil horizons were found including a bone bed. The records and fossils were given to Dudley Museum expanding its Carboniferous Coal Measures collection and providing the most detailed record held on a local site by the Geological Records Centre. The information was also used to recreate in exceptional detail a spectacular mining display of the Stinking Coal Seam (Worton 1994).

Case study 2

Engineering works can also be potentially damaging to the non-statutory protected urban geological resource. This occurs as a result of ignorance on the part of one or more of the parties involved. Stores Cavern on Dudley's Castle Hill provides an example of where this problem was narrowly avoided. In 1986 Stores Cavern yielded wonderful examples of the Wenlock trilobite *Trimerurs delphinocephalus* during the mining of the limestone on Castle Hill. These fossils were found in life positions and were not recorded elsewhere in the local area. The opportunity to make this find was almost lost. Some time after works had begun to stabilize this cavern for use by tourists, the museum was made aware of the works by chance through an employee of Dudley Zoo which is located on the hill. Once informed about the geological potential of the site the mines manager invited the museum's geological curator to visit the site and information was passed to the Black Country Geological Society and other local amateur geologists. These local geologists were able to extract over 20 tonnes of material from fossil bearing horizons at this site and it was this material which yielded the first *Trimerus* to be found for 130 years (Reid 1994). Ignorance of this development work almost caused the loss of this important site.

Case study 3

Engineering works may also be of a temporary nature pending the full implementation of a major scheme. This situation may allow lines of communication to get crossed or fouled-up. Hayes Cutting, Lye, West

Midlands provides a good example. In 1992 a section of retaining wall on the Hayes Road had deteriorated to such an extent that Dudley Metropolitan Borough Council proposed its removal and the excavation and regrading of the soil and soft rock behind. Because of its proximity to the historically famous Hayes Cutting, the Black Country Geological Society was consulted and given the opportunity to log a section of the soft rocks immediately above the Basal Carboniferous Coal Measure conglomerate exposed here. This provided a section previously unseen in recent times in this area. As part of this work the council provided a skip and some tree cutting work to clear the Hayes Cutting overgrowth (Cutler & Worton 1992). This work returned the cutting to a state approaching its former glory when it was first described by Murchison (1839). In 1994 it became apparent, through the author's employer, that works on the Lye by-pass would remove the cutting altogether through regrading and planting following a scheme to widen the road. Fortunately, the author was able to enter discussions prior to the finalization of the engineering design and recommendations for a new cutting to be constructed, *c*. 10 m back from the current section, were made. This will provide opportunities for interested parties to log and sample the section with the agreement of the site engineers. This compromise would not have been possible if the lines of communication had been opened too late, when details of the works had been finalized. The earlier geologists get involved the better.

The future of urban engineering opportunities

These case studies illustrate both the potential and some of the pitfalls of urban geological opportunities. In particular, they emphasize the problems of communication between the various parties involved. In a recent paper McCall (1994) noted that in all populated continents rapid urbanization is occurring, requiring *c*. 25 tonnes of earth materials per person per year. This means that there must be an expansion or major re-structuring of our urban areas to accommodate this. There is, therefore, no foreseeable decrease in engineering works and there will be no decrease in opportunities offered for urban geology by this industry. We need to examine how we might make the most of this opportunity and conserve the resource it produces.

Conclusions

Development as the geologist's tool

Engineering in urban areas can offer: near perfect exposure on the doorstep; fresh exposure of soft rocks and deposits which are otherwise easily weathered and difficult to maintain; and on-site engineering staff trained in sampling and recording and who may be sympathetic to geological needs.

To realize this potential the following are required: the involvment of appropriately trained or knowledgeable geologists at the very heart of the development process, from initial idea to the design and implementation of project; education of all those involved in development about the importance of geology; the respect of the wishes, codes of practice, instructions and working methods of the planners, engineers and the client particularly when on-site; the realization that most geological activities on engineering sites are the result of goodwill; and taking every opportunity to involve the site engineers and explain the importance of your work.

In practice, this means three things which geologists are not very good at. Firstly, we must provide the development community with clear statements of what we expect of them, what obligations they have to us and the advantages or potential consequences of their actions. Secondly, we must be involved in local planning policy and familiarize ourselves with its range of controls and laws. Thirdly, we must campaign to get geologists appointed in a planning/development advisory capacity in our local authorities to ensure that sites, established or new, are monitored and get planning conditions attached to any planning permission which may affect them.

Urban geological dreaming

Imagine a situation where we, as geologists, have a significant part in the decision making process and provide the engineers and planners with our specifications of things that we desire. Through influencing their policies and getting planning conditions into development documentation we may be able to do this. Why not have buildings designed to reflect our local geological heritage, sited sensitively in areas or sites of local

Fig. 12.4. Steve Field's vision of the proposed Wrens Nest geological centre. (Reproduced with kind permission of Steve Field, Borough Artist, Dudley Metropolitan Borough Council).

geological importance, with geology interpreted within and around as conditions for development (Fig. 12.4)? Why not be ready to advise when the future urban centres have to be created? Why not have a development control system which considers geology from the cutting of the first sod of the future new towns? Why not start now, getting to know the planners, engineers and developers in your local area and make it happen?

Acknowledgements

I would like to thank Dudley Metropolitan Borough Council; Mr Peter Mills, Mr Graham Bartlett and Mr Stephen Weston of Johnson Poole & Bloomer for their assistance in the Dibdale/Burton Road Project; Mr Vic Smallshire of the Dudley Canal Trust; The Black Country Geological Society and Dudley Museum and Art Gallery for their support for this paper.

References

Cutler, A. & Worton, G. 1992. The Hayes Cutting gets a facelift. *Earth Science Conservation*, **31**, 27-28.

McCall, J. 1994. Geoscience and the Urban Environment in Developing Countries. *Geoscientist*, **4**, 28-31.

Murchison, R. 1839. *Siluria*. London

Reid, C. G. R. 1994. Conservation, communication and the GIS: an urban case study. *In:* O'Halloran, D., Green, C. Harley, M., Stanley, M. & Knill, J. (eds) *Geological and landscape conservation*. Geological Society, London, 365-369.

Worton G. 1994. A person on the inside: opportunities for geological conservation in local engineering projects. *In:* O'Halloran, D., Green, C., Harley, M., Stanley, M. & Knill, J. (eds) *Geological and landscape conservation.* Geological Society, London, 359-363.

13 Opportunity docks: a case study of earth science conservation in an urban development

George R. Fenwick & Steven G. McLean

Summary

- A tufa dome was discovered during pre-development clearance work in a derelict harbour area of the River Tyne in Sunderland.
- This contribution documents the conservation of this dome and the methods used to stabilize and conserve it as part of the Marine Activity Centre planned for the area.
- It illustrates the conservation gain which can be achieved through the co-operation of local geologists and developers.

In 1992, Tyne and Wear Development Corporation began clearance and demolition work at Sunderland's North Dock as part of the St Peter's Riverside regeneration scheme. At the northwest corner of the dock, in front of and above the original nineteenth century stone harbour retaining wall, a number of old wooden huts were demolished in order to build a new Marina Activity Centre. In July of the same year the Corporation, through the contracted site engineers, contacted Sunderland University and Sunderland Museum to report that they had discovered a rather unusual feature which might be of interest to local geologists. Work on this particular section was halted until the authors were able to arrange a site visit. On arrival the feature was identified as a large calcareous tufa dome attached to the stone retaining wall between two stone buttresses (Fig. 13.1).

Calcareous tufa is the general term given to deposits of calcium carbonate ($CaCO_3$) formed by precipitation from aqueous solutions of calcium bicarbonate ($Ca(HCO_2)$). The North Dock tufa forms an overhanging dome which is approximately 6 m high, 5 m wide and has grown outwards from the wall to a depth of about 2 m. It is estimated that the entire structure weighs *c.* 80 tonnes. The tufa consists of a mass of plant debris enmeshed in a cellular calcium carbonate matrix. Over a

Fig. 13.1. The North Dock tufa as it was found in 1992.

period of time the tufa has grown, petrifying all solid matter attached to the wall, to create the present overhanging growth along with a series of stalactites and stalagmites. Periodic interruptions in the water supply probably allowed vegetation to grow over the surface and become petrified by precipitation as the water flow resumed.

After considerable discussion between the authors, who at this stage were also representing the interests of the local Regionally Important Geological/Geomorphological Sites (RIGS) group, Tyne and Wear Development Corporation and the site engineers, together with representatives from English Nature, all agreed that the tufa merited preservation as an outstanding local example of its kind. The site engineers were then commissioned to undertake a feasibility study into the methods by which the site could be structurally stabilized and incorporated into the fabric of the Marina Activity Centre. The subsequent report outlined in detail a number of engineering options which would have to be implemented to ensure the long term preservation of the feature.

Site conservation

The original plans for the Marina Activity Centre called for the demolition of the old harbour retaining wall and stone buttresses which would have necessitated the destruction of the tufa dome. Once it was agreed, however, that the tufa should be preserved, the designs for the Marina Activity Centre were architecturally modified so that the tufa could be kept as an integral feature of the site. The relevant section of the harbour wall and adjacent stone buttresses, together with the tufa dome, were to be retained and housed in an enclosed courtyard, open to the elements, within the entrance area of the complex (Fig. 13.2). It was recognized from the outset that extensive site stabilization work would create a degree of artificiality. This was, however, considered necessary and inevitable if the tufa dome was to be retained.

The structure and nature of the tufa gave rise to a number of technical problems. In practice it was very brittle having a cellular structure of laminated calcite with a high proportion of voids. In addition, the tufa dome itself was only attached to the face of the harbour wall and could have collapsed at any time. These factors, together with its weight, not only made conservation work difficult but created potential safety problems. Any stabilization work would result in a certain amount of

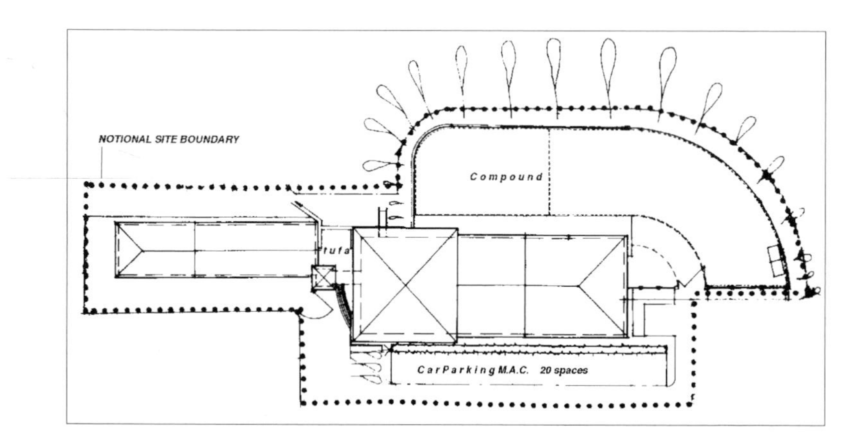

Fig. 13.2. Plan drawing of the Marina Activity Centre showing the location of the courtyard containing the tufa dome. Reproduced with permission of Tyne and Wear Development Corporation and Bullen and Partners (Consultant Engineers).

damage to the tufa but as it would naturally regenerate with time this was accepted, provided the damage could be kept to a minimum. Security of the site during stabilization and construction work was also of considerable importance. The tufa, even when stabilized, would be very fragile and would have to be protected from random sampling and malicious vandalism.

Engineering works

The first stage was to locate and control the water supply to the site. The construction area consisted mainly of glacial till overlying Upper Permian Roker Dolomite, but site investigation revealed that to the rear of the old harbour wall a small valley had been cut through the till into the bedrock beneath. This valley was filled with glaciofluvial sands. Groundwater, highly charged with calcium carbonate, was being channelled through this permeable buried valley to the rear of the wall, before flowing over and seeping through the large sandstone blocks of the wall. The water was traced back to an old railway bridge below Harbour View Road. At this point it was intercepted and temporarily diverted away from the site during construction work. After completion of the work, the water flow was reestablished so that continued growth of the tufa could occur in order to 'heal' any damage resulting from stabilization work. A control valve was located upstream of the site, and the water conducted to the rear of the wall where it was allowed to percolate through to the tufa dome. The water was then to be drained away into the North Dock via a pool constructed at the base of the tufa dome.

The second stage of the work involved the removal of loose tufa and the erection of temporary supports for the dome. Loose tufa, underneath the main dome, was removed using a JCB with a long reaching back-hoe. The tufa removed was stored for later sampling. Soil and fill were then extracted to a depth of *c.* 1 m to allow for a concrete foundation (Fig. 13.3). This was to provide a stable platform on which to erect for the temporary supporting structures. A number of designs for the temporary supports were considered and the final system adopted was chosen as the one which provided the maximum support with the minimum damage. In this system the tufa was supported by two steel beams angled at about 45° from the concrete foundation. These were

Fig. 13.3. Construction of the upper section of the concrete foundation and installation of hydraulic supports.

drilled into the front edge of the dome. A number of temporary hydraulic props were also installed to support the dome from below (Fig. 13.3).

The third stage of the operation involved the bolting of the dome to the retaining wall to provide permanent stablization. First, the sand and clay behind the wall were excavated to expose the rear of the wall, and a scaffolding drilling platform constructed (Fig. 13.4). The wall and tufa was then core-drilled so that a system of stitch anchors, bolted to steel plates at the rear of the wall, could be installed (Fig. 13.5). During the drilling operation a number of core samples were collected. These are stored at Sunderland Museum and await future research. Epoxy grout was then forced through socks surrounding the stitch anchors to emplace them securely within the tufa. As a further precaution against corrosion the steel plates and bolts to the rear of the wall were coated in a bitumen paint. This work was carried out by a specialist drilling company. The final permanent support was then installed. This was in the form of a pre-cast, steel reinforced, concrete frame, consisting of a horizontal beam supported by two uprights securely bolted and embedded into the upper and outermost sections of the concrete platform below. This was designed to carry the main weight of the dome. To ensure maximum support, the upper horizontal member of the frame had to be as closely fitted to the underside of the dome as possible. A groove was therefore cut into the underside of the dome to house this support. Although this frame is

Fig. 13.4. Scaffolding platform to rear of the retaining wall during drilling operations.

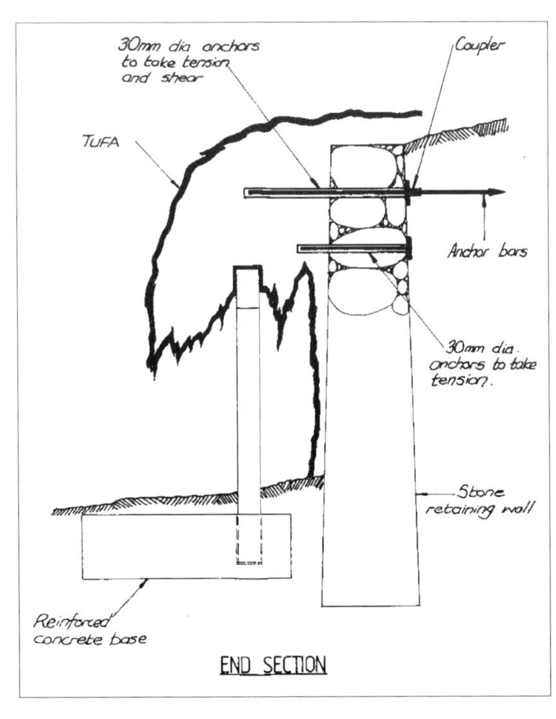

Fig. 13.5. Cross-section of retaining wall and Tufa showing stitch anchorage and pre-cast concrete frame. Reproduced with permission of Tyne and Wear Development Corporation and Bullen and Partners (Consultant Engineers).

quite visible and is obviously not a natural part of the tufa dome it was essential for both structural and safety reasons. In an attempt to enhance the future appearance of the site, water has been allowed to flow down the frame in the hope that ultimately the frame itself will become calcreted. To aid this, the surface of the concrete was allowed to remain rough to help facilitate a build-up of new tufa deposits. Once the dome was firmly supported from below, and securely bolted to the wall, the temporary supports to the front and base were removed. The scaffolding to the rear of the wall was dismantled and the excavation refilled with the original permeable sands.

The final stage of the operation involved the construction of the Marina Activity Centre (Fig. 13.6). During this period, a small shallow pond was constructed at the base of the tufa and the drainage system completed. Smooth plastic pipes were used in order to reduce the amount of tufa build up within the drains. Damaged sections of the old harbour retaining wall were repaired with new stones cut specifically for the purpose, and the butresses were repointed. Steps from the building and a viewing platform were also constructed, and finally lighting was installed in the courtyard of the centre in which the dome was located so that the tufa could be viewed at night. Work on the site was completed during the summer of 1994 (Fig. 13.7). It is hoped that some interpretative panels will be installed around the pool in the near future.

Fig. 13.6. The Sunderland North Dock Marina Activity Centre in January 1995.

Fig. 13.7. The North Dock tufa after completion of site works.

Maintenance and site status

The site requires periodic inspection and maintenance. A programme of maintenance has been arranged with Tyne and Wear Development Corporation and this responsibility will subsequently pass to Sunderland City Council when the Corporation completes its work on the riverside in 1997. At its simplest, this involves clearing the pool of leaves and rubbish, controlling the vegetation on the tufa dome itself, and checking that the drainage pipes have not scaled up. The site will also be monitored for any damage and vandalism. More importantly, however, the maintenance programme also involves regular checks on the stability of the tufa dome and the condition of the supporting frame to ensure that any deterioration is noted.

It was recognized at an early stage that the site is one of the best examples of its kind in the northeast of England and merited conservation on these grounds alone. For this reason, a dialogue was opened between the local RIGS group (RIGS North East) and the planning department of Sunderland City Council. It was recommended that the site be protected through the City Council's Unitary Development Plan as a RIGS. A proposal to the council was drawn up by the RIGS group, through consultation with Tyne and Wear Development Corporation, and presented to the City Council during the autumn of 1994. The site will be presented to committee where it is hoped that the proposal will be successful.

Publicity has been deliberately kept to a minimum because throughout most of the period when the stabilization and construction works occured, the site was off-limits to the public on safety grounds. There was also a policy of non-disclosure until the site was secured within the courtyard because much of the time it was only temporarily fenced, and damage could have occurred through vandalism. Since completion, the site has had some publicity through BBC Radio Newcastle, but is to be formally launched when confirmation of its RIGS designation has been received.

Conclusion

The Sunderland North Dock Marina Activity Centre now hosts a feature of geological interest within its walls and one which will, it is hoped, be designated as a RIGS. It is also hoped that the tufa dome will develop into a local attraction and educational facility and thereby promote the concept of geological conservation. The site is accessible to anyone, and should remain safe because access to it is strictly controlled through the building. The scientific and educational value of the tufa dome has been preserved despite the fact that a great deal of invasive work has been undertaken. A complete photographic record was made throughout the two years of development. In addition, water samples were collected together with cores and loose tufa samples all of which have been lodged with local public institutions. The conservation of the site is now complete, but the scientific investigation continues.

In the course of major land development, geological or geomorphological features are all too often damaged or destroyed. This is particularly true in the urban environment where the pressures of land development are high. The Sunderland North Dock tufa site demonstrates that the often conflicting goals of land development and site conservation can be resolved, given a sympathetic developer, local geological support, and co-operation from the local planning department. Tyne and Wear Development Corporation has supported the conservation of the site from the onset, providing access to a considerable number of resources, not least financial.

The complex problems of conserving this particular feature were resolved by using experts in engineering, drilling, construction and architectural design. At the same time, this exercise is a model of what can be achieved through cooperation between geologists and developers. Provided developers are willing to inform local geologists of their discoveries, that 'risk' can, as has been shown at this site, turn into a triumph for all.

Acknowledgements

We acknowledge the help and assistance of the following organizations: Tyne and Wear Development Corporation, Sunderland City Council, English Nature (Northumbria Team), Bullen and Partners (Consulting Engineers), The Alan. J. Smith Partnership; and to the following individuals: Mr A. James, Mr M. Agar, Mr A. Coles and Mr L. Golding (photography), and to all those who worked on the site, without whose help this project would not have been possible.

Part Three

Awareness and use of the urban geological resource

Awareness of the value of earth heritage and the need for its protection are some of the most important aspects of the urban environment, where over 80% of the population of the United Kingdom live. There are three sections to this part. The first, ***The role of the local authority***, demonstrates the pivotal role of the local authority in promoting awareness and in protection of vulnerable urban sites. The second, ***Urban geology and education,*** is illustrative of the value of urban geology in education. The third section, ***Increasing public awareness and involvement***, gives practical guidance in the promotion of urban geology through local community involvement, through interpretative signing, and through the media.

14 Local geology and local authorities

Ed A. Jarzembowski

Summary

- Fieldwork is fundamental to the study of geology and, often, local societies have paved the way in documenting the nature and scope of the urban resource of exposed geology.
- There is a clear need for an environmental audit of regional geological resources at county and district level.
- This contribution looks generally at the role of local authorities within urban geological conservation and considers how local geological groups can best exploit this.

Local geology

It is often said that the best geologist is the one who has seen most rocks. This implies that to be a good geologist one must travel the globe looking at exotic rocks. In practice, however, the great majority has neither the resources nor the inclination for such expeditions. Most of us live in built-up areas where the usual 'rocks' are bricks and mortar, concrete and tar macdam. Nevertheless, even as an impecunious youth in London, I found that there was plenty of opportunity to enjoy geology. Books could be obtained from the local library, the formerly specimen-laden public museums in South Kensington could be visited and the weathered stones on historic buildings and in open spaces could be examined. More importantly, there was the opportunity to collect on building sites where deep excavations exposed London Clay.

In time, it was relatively easy to assemble a collection of Lower Eocene (London Clay) fossils from such temporary exposures. A trip on the underground would take you to more permanent exposures such as the old chalk pit at Harefield or tidal outcrops on the Thames foreshore in

London. The inevitable pyritic decay of specimens encouraged me to seek advice from experts in the Geology Section of the London Natural History Society and the Tertiary Research Group. Occasionally, more distant excursions with the family and even the Geologists' Association provided the opportunity to diversify the collection, and friends and relatives would often add items too. The collection became a source of discussion as well as providing material for school projects and displays. The down side was that very few teachers had any inkling of geology, and so universities required no 'O' or 'A' levels in the subject to read earth science. Most schools have neglected the subject and the great majority of the United Kingdom adult population has therefore never received any geological education. Understanding is inevitably low and, at best, geology is confused with archaeology or included within geography.

The National Curriculum and adult education are beginning to change this situation. Nevertheless, geological organizations and museums have an invaluable role to play in raising awareness and in supporting teachers through the provision of advice and materials. Geological societies should be encouraged to assemble local information on sites and specimens. For many, urban localities have the advantage of shorter journey and visiting times, with facilities such as refreshments and conveniences being not too far away. Buildings and cemeteries provide varied 'outcrops', and there are additional opportunities to look and learn in coastal towns with cliff sections, or in towns adjacent to open landscapes. However, opportunities to collect from commercial excavations are often beset with dangers. Fossil or mineral sites in, or near, urban areas and where hammering is unnecessary, are vital (e.g. former colliery tips). Often being slightly further away, they may be ideal for organised outings or treats in school holidays with parents in attendance. Extra rewards, such as a certificate, help make the day more memorable.

Nature conservation trusts were previously preoccupied with wildlife in the countryside, but are now beginning to develop interest in the urban scene and in geology too, including expanding the WATCH scheme for young people (RockWATCH). Such trusts, like amenity societies, often receive some local authority support and there is much more that a local authority can do.

Local authorities

The two major areas where local authorities have a key role in promoting geological conservation are through development planning and leisure services.

Development planning

Liaison with planning officers in local authorities is essential to ensure that the public value of geological sites is recognized in land use documents and consultations (Table 14.1). With regard to development planning, local authorities have a central role in safeguarding earth science sites through the planning system. Not only do they have a role in development control, handling all planning applications that may damage or even enhance sites, they are also responsible for production of local plans which identify all areas of nature conservation interest. Therefore, if a RIGS group has worked closely with sympathetic local planners, there is every chance that RIGS sites will join SSSI on the local plan map. This ensures, at the very least, that the earth science interest of the site is considered as part of any planning application. In any case, liaison with planners can provide information on a diverse range of new developments, some with geological opportunities. In addition, minerals officers will know of quarry and pit developments, and highways officers of potential roads and tunnelling sections. Officers dealing with waste disposal will be aware of sites identified for landfill, and archaeologists are likely to be aware of, and have access to, many of the above mentioned excavations.

The publication, late in 1994, of *Planning Policy Guidance: Nature Conservation* (*PPG 9*) by the Department of the Environment will further safeguard the natural environment through encouraging development control decisions that respect the natural enviroment, and through raising the awareness of nature conservation, including specific mention of earth science conservation, within local authorities.

Local authorities also have the power to designate sites as Local Nature Reserves, and are also required, as a follow up the 'Earth Summit' in Rio de Janeiro, to produce a Local Agenda 21, which involves the local community and local business in setting environmental standards for the local environment. It is worth knowing the priorities of the local politicians

Table 14.1. Extracts from Brighton Council's *Charter for the Environment* (1990) with policies for geological conservation and site management.

- There will be a presumption against any development taking place which will affect the Sites of Special Scientific Interest and Local Nature Reserves in the Borough and the Council will support any management schemes necessary to secure the continued protection of these areas.
- The Coucil will take account of ecological and geological considerations in implementing its own programmes and in determining proposals for development in order to safeguard ecologically and geologically sensitive areas and to enhance nature conservation. Consent will not normally be given for the development of important flora and fauna habitats and geological sites. Where development is to take place which affects such sites, conditions or agreements will be imposed on planning permission to allow for geological investigations to take place before and during development as may be appropiate and developers may be approached for funding.

Note: There has been comparatively little take-up of the policy; indeed, there is only one RIGS identified so far in the whole of Sussex. This emphasizes the need for geological participation as well as policy formulation.

(the Councillors), however, in case lobbying on certain issues is required.

With the proper steer, all of the above activities offer a great opportunity for enhanced conservation of geological and geomorphological features in the urban environment.

Leisure activities

Many local authorities now provide informal education and activities outside working hours as part of their leisure services through countryside rangers. This includes guided walks as well as practical work including the provision of sign posting and information boards at strategic points. Geological organizations can do well to liaise with such services, and offer to lead walks and support generally.

To support such activities, a local register of sites suitable for education and research is needed, as well as up-to-date information on conditions for visits, and what may be seen and found. This can be readily undertaken by working with local museums which are often maintained by local authorities as part of their leisure provision. Geological societies and museums can combine to provide materials for evening classes and such activity may even help to preserve local museums in times of financial cut-backs. Although museum curators may be tied to their collections, site management can be pursued by rangers as part of practical conservation schemes. Many local authorities employ regional archaeologists who may provide skilled help with excavation of larger finds, such as vertebrates. There is no denying that local authority funded museums, wardens and rangers, and the activities they undertake, have an important role in promoting geology and its conservation

Conclusion

For local geology to be more effective it must receive some resourcing. Whilst professional geologists actively seek out engineering contracts, collecting and basic research is less well organized and supported. It is therefore worth pursuing geological opportunities as part of planning gain and/or applying to the council for public grants towards projects or individual items (e.g. sign posts). There is a continuous need to keep up the profile of geology, ranging from 'nature' and 'events' columns in the

Table 14.2. Opportunities for collaboration between local geologists and local authorities.

Services	Topics
Archaeology	Excavation assistance
Ecology	Geological nature conservation
Education	Site/specimen information
Environment Forum	Sustainable development/Local Agenda 21
Grants	Resourcing
Landscape	Geological and geomorphological features
Minerals	Extraction sites
Museums	Site/specimens databases/displays/public lectures
Open spaces	Geological trails
Planning	Site protection/investigation
Promotion	Publicity
Rangers	Site management/interpretation
Transport	Geological sections
Waste	Landfill sites

local press to announcing finds on the local news. Many local authorities now have officers dealing with press and media matters, as well as publishing free newspapers and programmes of activities. Participation in fairs and festivals is often encouraged. These all provide suitable 'vehicles' for spreading the word, especially if the human interest factor is taken into account. It is important to talk to non-geological as well as geological audiences, including other organizations such as the Womens' Institute, and even offer presentations to planners and newly elected councillors. Table 14.2 provides a brief summary of local authority services and corresponding geological interests. Finally, do not forget to emphasize the fun aspect of geology, which can help to break down barriers, and which probably attracted us to the science in the first place.

References

Department of the Environment. 1994. *Planning Policy Guidance: Nature Conservation. PPG9*. HMSO, London.

15 A code of practice for geology and development in the urban environment

Colin Reid

Summary

- In heavily urbanized areas geological exposures are often rare and transient.
- There is a constant flux in which new development continually creates temporary exposures whilst threatening existing geological sites.
- The ability to monitor this development, or better, to control it in a positive way is the key to success in urban conservation.
- Within the Metropolitan Borough of Dudley a newly adopted Code of Practice aims to make this possible and to improve the interaction between the geologist and developer.

In recent years the significance of geological exposures in the urban environment, and the need to monitor and conserve these, have been promoted by many within the geological community (Prosser & Larwood 1994). In contrast to rural areas, the urban environment is one of constant activity in which development is continually changing the face of the landscape. Consequently, geological sites in our towns and cities tend to be localized, and are often short-lived.

Urban development undertaken without consideration for these rare and often very important geological features can damage or destroy them. Conversely, development can also create new exposures, both temporary and permanent, and the opportunity to use these for research, education or recreation. This is well illustrated in the case of Dudley in the West Midlands, which has over 200 sites including six Sites of Special Scientific Interest (SSSI), and 25 'second tier' sites known in the West Midlands as Sites of Importance for Nature Conservation (SINC). The majority of both types of site were created as a direct result of human activity; notably road, rail and canal cuttings, and active or disused mines and quarries.

The problem facing conservationists in Dudley has been how to monitor potential development so that opportunities to preserve or record sites can be seized when they arise. This can only be done effectively by keeping a finger on the pulse of local planning activities.

In its role as a Geological Recording Centre (GRC) for the Black Country, Dudley Museum has been in an ideal position to carry out this monitoring task, by using channels within the local authority system to vet new planning applications, existing planning permissions and any relevant works being carried out by the authority itself. A computerized Geographical Information System (GIS) pioneered by Dudley has proved to be an invaluable new tool in this process (Reid 1994). However, the recent merger of the authority's Planning and Leisure departments, the latter having responsibility for the Museum, has allowed the GRC to take a more proactive role by directly influencing planning policy to accommodate conservation issues.

The outcome of this is a radical new code of practice for developers (Dudley Metropolitan Borough Council 1995), designed to protect Dudley's unique geological heritage and, where possible, to enhance this resource. The code has been produced in the wake of Dudley's Unitary Development Plan (Anon1994) and of the Black Country Nature Conservation Strategy (1994) in which Dudley and the other Black Country boroughs (Wolverhampton, Walsall and Sandwell) commit themselves to a range of policies to preserve, protect, promote and enhance the area's geological and biological heritage (Cutler 1996). Both of these documents complement national planning policy as expressed in the Department of the Environment's (1994*a*, *b*) Planning and Policy Guidance 9 (PPG9) - Nature Conservation, and PPG12 - Development Plans and Regional Planning Guidance, together with the Wildlife and Countryside Act 1981.

The Code of Practice

The Code sets out procedures to be followed wherever development proposals affect existing geological features, or where they are likely to expose fresh sections through the bedrock. It is relevant to all those involved in undertaking, facilitating or promoting new development, council departments and officers, statutory undertakers (such as gas, water and electricity companies), government departments and agencies, and private developers.

Produced in the form of a glossy A5 pamphlet the Code is intended to be made available to all those submitting planning applications. In its introduction the Code stresses the importance of the local geology, the need to respect this and the positive role developers have to play in the conservation process through partnership with the local authority.

The key aim of the Code is to ensure early consultation with the borough geologist so that conservation issues are considered prior to the decision-making process on planning applications, rather than at a late stage, or not at all, as has previously been the case. While the procedures laid out are mandatory, the authority's intention is to apply the code sensitively and not snarl up the planning system with additional and unpopular bureaucracy. Indeed, the vast majority of planning proposals will be unaffected by the code.

The Code of Practice will help to preserve and enhance Dudley's geological resource by promoting the following. (1) Conservation of existing designated geological sites. This will always be sought as a first option, and can be achieved by the sensitive design of new development. Planning permission may be refused where development proposals would result in unacceptable damage to or loss of important geological sites or features. (2) Conservation by record of temporary exposures (i.e. through sampling and recording). This will be required through planning condition or legal agreement where physical conservation of the resource *in situ* cannot reasonably be achieved. (3) Creation of new sites, either of a temporary or permanent nature, to use the opportunities provided by development.

The Code identifies a wide range of activities that can affect existing geological sites or create new ones. These comprise: road schemes, rail links, canals and related infrastructure; water, gas and electricity infrastructure; new building development; derelict land reclamation schemes and landscaping; mineral extraction/quarrying; landfill operations/tipping and discharge of any materials; agricultural and forestry activities, including planting and afforestation and grading or seeding of rock faces; council or other environmental improvement schemes of all types; council plans and policy formulation affecting geological sites and surface or sub-surface geological features; council highways, building schemes/structural engineering services, including those affecting mines and caverns; council land and building acquisition, management and sale; council economic development activities; tourism related development of geological sites; unsolicited removal of geological material from sites, including illicit commercial collecting or over collecting; unsolicited use of foreign hardcore on or adjacent to geological sites.

The Code requires that developers:

- recognize their obligation to assess the geological implications of development proposals at the earliest possible stage and provide full information to the planning authority as to the likely impact of the proposals.
- recognize that consultation of the sites register in the Geological Recording Centre at Dudley Museum is the most effective method of accessing geological information.
- recognize that effective early consultations must have taken place between the borough geologist and prospective developers before schemes affecting geological features are offered for planning approval or brought towards implementation.
- recognize that it may be necessary for a full geological assessment of a development site to be undertaken prior to consideration of a planning application.
- be prepared to discuss mitigation of any impact the development may have on significant geological features.
- recognize that if the most appropriate course of action is deemed to be conservation by record only, i.e. sampling and recording, site access and time will need to be made available for this to take place, and that a contribution towards associated costs may also be requested.
- recognize that if it is deemed necessary to remove an existing geological feature in the course of development then, where appropriate a new/alternative feature should be created in its place, and the cost borne by the developer.
- recognize that the local authority can provide advice on all geological matters, and also: (1) provide expertise for site investigation and recording on behalf of developers if requested; (2) provide developers with details of geological contractors capable of carrying out necessary

recording work; (3) monitor geological work carried out for developers, to ensure compliance with the specification and completion of the work to the satisfaction of the local planning authority.

Applying the Code of Practice to planning application on existing designated sites

Following an initial enquiry by the developer, a geological assessment of the site would be made by means of a desk-top study, based on existing information held by the Geological Recording Centre. If the features under threat are deemed to be significant then discussions would be held on the potential impact of development and possible avoidance or mitigation of damage to the site. Depending on the outcome of these discussions and subsequent action, a number of outcomes is possible.

- Planning permission is refused because the significance of the geological features outweighs the need for damaging development.
- Planning permission is granted as development has been designed to avoid damage to the geological features.
- Planning permission is granted on condition that an alternative geological site is created to replace that destroyed by the development.
- Planning permission is granted with conditions/binding agreement that access and time are made available before/during development for recording and sampling to mitigate unavoidable damage.
- Planning permission is granted with conditions allowing the borough geologist a watching brief during development.

Applying the Code of Practice to planning applications on potential new sites

Again, the prospective developer's enquiry would be followed up by an assessment of the potential significance of new exposure on the site, based on a desk-top study of the local geology. Such a study by itself is unlikely to prove entirely satisfactory, however, as the true potential of the site would not be known until the exposure or exposures have actually been created. If it is felt that significant features might be revealed, discussions would be held on the site's potential, and as how this might be accessed and used to the satisfaction of all parties.

As before a range of outcomes is possible as a result of assessment and discussion.

- The borough geologist may request a watching brief on the development as a condition of planning permission.
- In the case of exposure being temporary, planning permission is granted with conditions/binding agreement that access and time are made available during development for sampling and recording.
- Planning permission is granted with development designed to create new permanent exposure(s).

The full range of scenarios outlined above is summarized in the accompanying flow chart (Fig. 15.1). Normally, development revealing fresh exposure would be encouraged as it would, at the very least, contribute to knowledge of the local geology. Exceptions to this rule might be where the development would destroy a buried feature of importance or where it would damage the nature conservation or other heritage value of the site.

Conclusion

The sensitive application of Dudley's new Code of Practice promises to reap many benefits. It will provide an effective early warning system to protect the Borough's existing geological sites. It will build a positive relationship between geologist and developer with local and scientific communities benefiting from this cooperation. It will create proper lines of communication on conservation issues within the local authority itself, which have previously been tenuous at best. It will give geologists the chance to seize opportunities that have too often gone begging in the past. Most importantly, the Code should overcome the great enemy of the urban conservationist, that of ignorance, and allow a degree of control over conserving the local geological heritage that has hitherto been impossible.

It is hoped that the example set by Dudley can be followed in other urban areas. If so, not only would the future of our precious urban sites be secure, but many new discoveries also lie ahead.

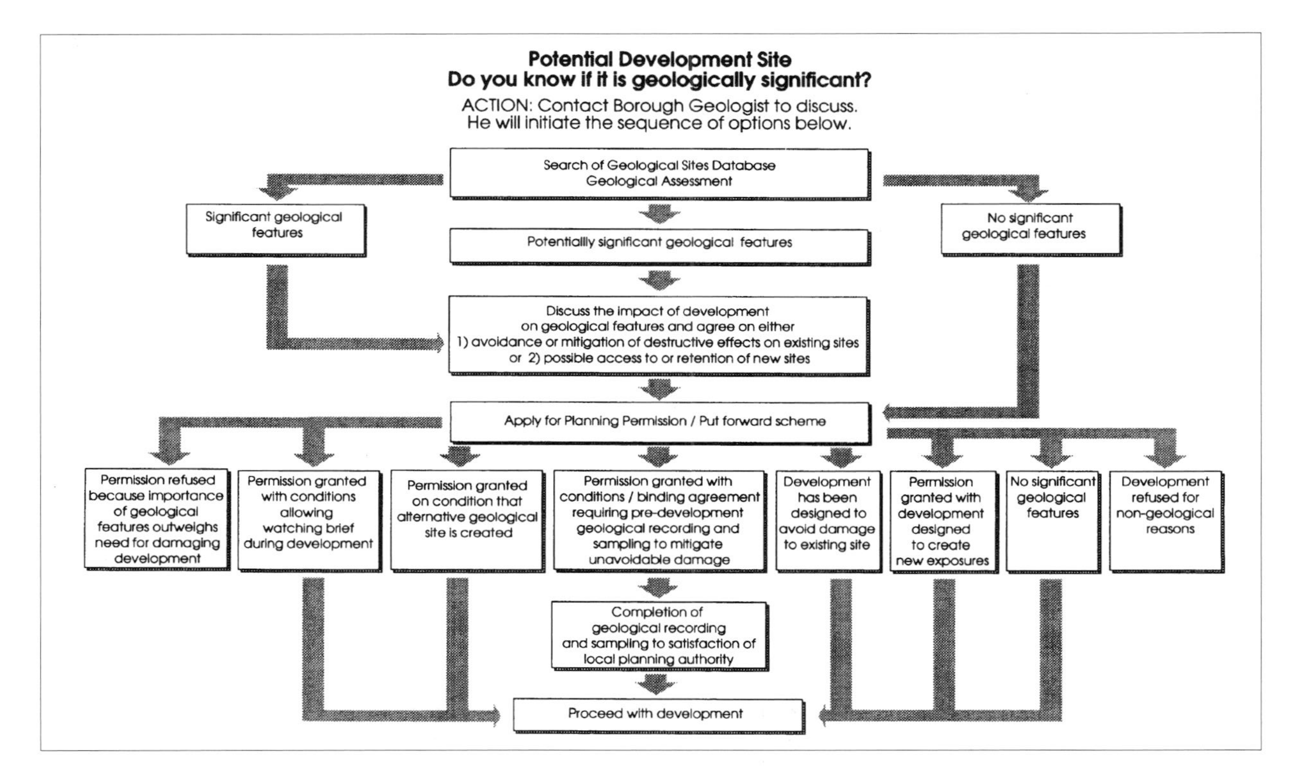

Fig. 15.1. Flow chart showing the application of the Code of Practice.

References

Anon. 1994. *Black Country Nature Conservation Strategy*. Compiled by: English Nature, Sandwell Metropolitan Borough County, Dudley Metropolitan Borough County, Walsall Metropolitan Borough County and Wolverhampton Metropolitan Borough County. English Nature, Peterborough.

Cutler, A. 1996. The role of the regional geological society in urban geological conservation. *This volume*.

Department of the Environment. 1994*a*. *Planning and Policy Guidance 9, Nature Conservation*. PPG 9, HMSO, London.

---- 1994*b*. *Planning and Policy Guidance 12 - Development Plans and Regional Planning Guidance*. PPG 12, HMSO, London.

Dudley Metropolitan Borough Council. 1994. *The Dudley Unitary Development Plan*. Dudley Metropolitan Borough County, Dudley.

---- 1995. *Geology and Development in Dudley: a code of practice for early consultation*. Dudley Metropolitan Borough County, Dudley.

Prosser,C. D. & Larwood, J. G. 1994. Urban site conservation - an area to build on? *In:* O'Halloran, D., Green, C., Harley, M., Stanley, M. & Knill, J. (eds) *Geological and Landscape Conservation*. Geological Society, London, 347-352.

Reid, C. G. R. 1994. Conservation, communication and the GIS: an urban case study. *In:* O'Halloran, D., Green, C., Harley, M., Stanley, M. & Knill, J. (eds) *Geological and Landscape Conservation*. Geological Society, London, 365-369.

16 Urban geology and the National Curriculum

Duncan Hawley

Summary

- All school pupils between the ages of five and sixteen are entitled to be taught some geology through the Science National Curriculum.
- The nature of this geology requirement is discussed and the problems associated with its delivery within the Science National Curriculum is considered.
- As a result, it is apparent that it is essential for local geologists to provide material within local areas for teachers who may have limited geological experience.

From September 1995 schools started to teach a revised version of the National Curriculum. For science this will be the third version since its introduction in 1989. The original version of the Science National Curriculum (Department for Education and Science 1989) contained in the Programme of Study a substantial proportion of geology, approximately 10%, together with a specific Attainment Target, 'The Earth and Atmosphere'. This target outlined the performance indicators to be used to assess a pupil's performance in learning within the earth sciences. This was significant since for the first time all pupils in England and Wales, between the ages of five and sixteen, would be taught a science course containing geology. An element which had not previously featured significantly or systematically in either primary or secondary school teaching. Earth science had gained a clear and distinctive profile in the curriculum.

The second version the Science National Curriculum (Department for Education and Science 1991) was a response to the unmanageable nature of the original version. Teachers had found that there was not enough time in the school day to cover all the content required and, in addition, to assess a pupil's performance. The inevitable rationalization led to a small reduction in the amount of curriculum content and a major

Table 16.1. A summary of the geology component of the Science National Curriculum (1995).

Key Stage One (5 to 7 years)
Grouping materials. Pupils should be taught:
- to sort materials into groups on the basis of simple properties, including texture, appearance, transparency and whether they are magnetic or non-magnetic;
- to recognize and name common types of material (e.g. metal, plastic, wood, paper, rock) and to know where some of these materials are found naturally.

Key Stage Two (8 to 11 years)
Grouping and classifying material. Pupils should be taught:
- to compare everyday materials (e.g. wood, rock, iron, aluminium, paper, polythene) on the basis of their properties including hardness, strength, flexibility and magnetic behaviour, and to relate these to the everyday uses of materials;
- to describe and group rocks and soils on the basis of characteristics including appearance, texture and permeability.

Key Stage Three (11 to 13 years)
Planning and experimental procedures. Pupils should be taught:
- to consider contexts where variables cannot readily be controlled (e.g. fieldwork), and how evidence may be collected in these contexts.

Changing materials: geological changes. Pupils should be taught:
- how rocks are weathered by expansion and contraction and by the freezing of water;
- that the rock cycle involves sedimentary, metamorphic and igneous processes;
- that rocks are classified as sedimentary, metamorphic or igneous on the basis of their processes of formation and that these processes affect their texture and the minerals they contain.

Key Stage Four (14 to 16 years) - Double Science
Planning and experimental procedures. Pupils should be taught:
- to consider contexts where variables cannot readily be controlled (e.g. fieldwork), and to make judgments about the amount of evidence needed in these contexts.

Changing materials: useful products from metal ores & rocks. Pupils should be taught:
- that a variety of useful substances can be made from rocks and minerals.

Changing materials: geological changes. Pupils should be taught:
- how igneous rocks are formed by the cooling of magma, sedimentary rocks by the deposition and consolidation of sediments, and metamorphic rocks by the action of heat and pressure on existing rocks;
- how the sequence of, and evidence for these processes are obtained from the rock record;
- how plate tectonic processes are involved in the formation, deformation and recycling of rocks.

change in the number of Attainment Targets: a reduction from 17 in the orginal version to just five. Earth science lost its distinct status and became a strand within the Attainment Target termed 'Materials and their Properties'. This resulted in geology being grouped with chemistry and in many secondary schools this resulted in the responsibility for teaching geology being allocated to chemistry specialists.

The latest version (Department for Education 1995), brought about by the review conducted by Sir Ron Dearing, has seen further changes, mainly at primary level. Here there has been further reduction in curriculum content, although the geological component is retained under 'Materials and their Properties'. This version is unlikely to be subject to further change since an assurance has been given to retain the curriuculum current form until at least the end of the millennium. Consequently all pupils between the ages of five and sixteen are entitled be taught about rocks. A summary of the geology component is given in Table 16.1.

The teachers

The great majority of teachers who are required to deliver the geological component of the National Curriculum have little or no expertise in geology. Most primary school teachers do not have any expertise or training in science, although confidence and competence in teaching science is increasing.

In secondary schools, who gains responsibility for teaching the geology component is dependent on how the science department is organized. In some schools it is shared out amongst subject specialists, but it is frequently delegated to the member of staff who is the chemistry specialist, because the geology component falls within the 'Materials and their Properties' Attainment Target, which is broadly seen as the Chemistry section of the National Curriculum. Many of these chemistry teachers do not feel comfortable with this responsibility. This is illustrated by the fact that when teachers were asked to send comments to the Dearing Review there was a significant number of chemistry teachers who suggested that the 'rocks' component should not be included in science and should be given over to Geography or removed altogether.

In consequence, many teachers' expertise in geology is based on 'book' knowledge, derived from general school-based texts, and there is a heavy reliance on the resource material available. For example, the demand for the 59 practical teaching units in the series 'Science of the Earth' produced by Earth Science Teachers' Association (ESTA 1988, 1991) has been great and over 46 000 have been distributed to date.

Teachers are also strongly influenced by the experiences they gain through In-service Training (INSET) courses. These increase teachers' confidence and help to frame the geological component of their lessons.

Many teachers are discouraged by the use of an array of technical geological terms and become uneasy when confronted with an array of complex names and labels. In the light of this general lack of geological confidence, there is an understandable preference towards using the minimum possible framework of terms. As a consequence, teachers find most useful a knowledge and understanding of how to identify rocks and distinguish between igneous, sedimentary and metamorphic types. Given confidence, time and enthusiasm, they will no doubt build up their geological vocabulary and knowledge to the levels they require.

In summary, the geology component of the Science National Curriculum is being delivered by teachers with little expertise in geology, and who are generally eager to gain advice on what they should be teaching and upon the best approaches to achieve this.

The teaching content

It is worth making a brief examination of what in the way of geology is required to be taught in the new Science National Curriculum in order to understand the teaching implications.

In Key Stage One, pupils are to be taught how to sort, recognize and name common materials, including rocks, on the basis of simple properties. In Key Stage Two, pupils are required to compare materials (rocks) on the basis of their properties, and to describe and then group them on the basis of characteristics. It is only in Key Stage Three that pupils are expected to be taught about classifying rocks into igneous, sedimentary and metamorphic types. This understanding should arise from relating the processes of rock formation to the appearance of the rock by asking: how did this rock acquire these characteristics?

There is a progression in learning here, starting with simple focused exploration and sorting using relatively open observation, moving through comparison and grouping, which involves the fixing of criteria, to the building of a general model which relates process and product, a model for which properties, common to each rock group, must be derived.

There is also a progression in terms of classificatory procedures. Forms of classification based upon the visual recognition of objects are appropriate to pupils of Key Stage One age, but it takes several years to move from the sorting of shapes and colours to the classifying based on more abstract groupings. Vocabulary is also one prerequisite for classification as well as being itself a product of classification. It is necessary to build up a vocabulary that will help to describe and distinguish before it is possible to generalize in.a meaningful way.

All this fits with the requirement to teach science through an experimental and investigative approach as outlined in a Programme of Study and which applies across all aspects the science curriculum.

The key message here is that the requirement to teach geology in the National Curriculum follows good teaching practice and places emphasis on developing an understanding of rocks using a systematic process based on observing the characteristics and properties of rocks. The significance of this interpretation is that the naming of rocks is seen as an end result rather than a starting point. Figure 16.1 shows an example of how this might be achieved for Key Stage Three pupils.

From the educational viewpoint it is important that pupils learn about rocks through systematic observation and investigation which is built upon their experience, rather than by didactic transmission of a catalogue of rock facts and names.

The urban geological environment as an educational resource

It is regarded as good practice in primary teaching to use, where possible, children's own experiences as a starting point for a study. Teachers in primary schools will therefore generally be keen to include some some sort of study of rocks in the local environment, in order to enable their pupils to experience and place rocks within a real context. A cross-curricular approach to teaching is often used in primary schools, although science is increasingly being taught as a separate subject at Key Stage Two. The 'project' is likely to be the vehicle by which rocks will be studied. A survey of primary schools in Gloucestershire in 1991-92

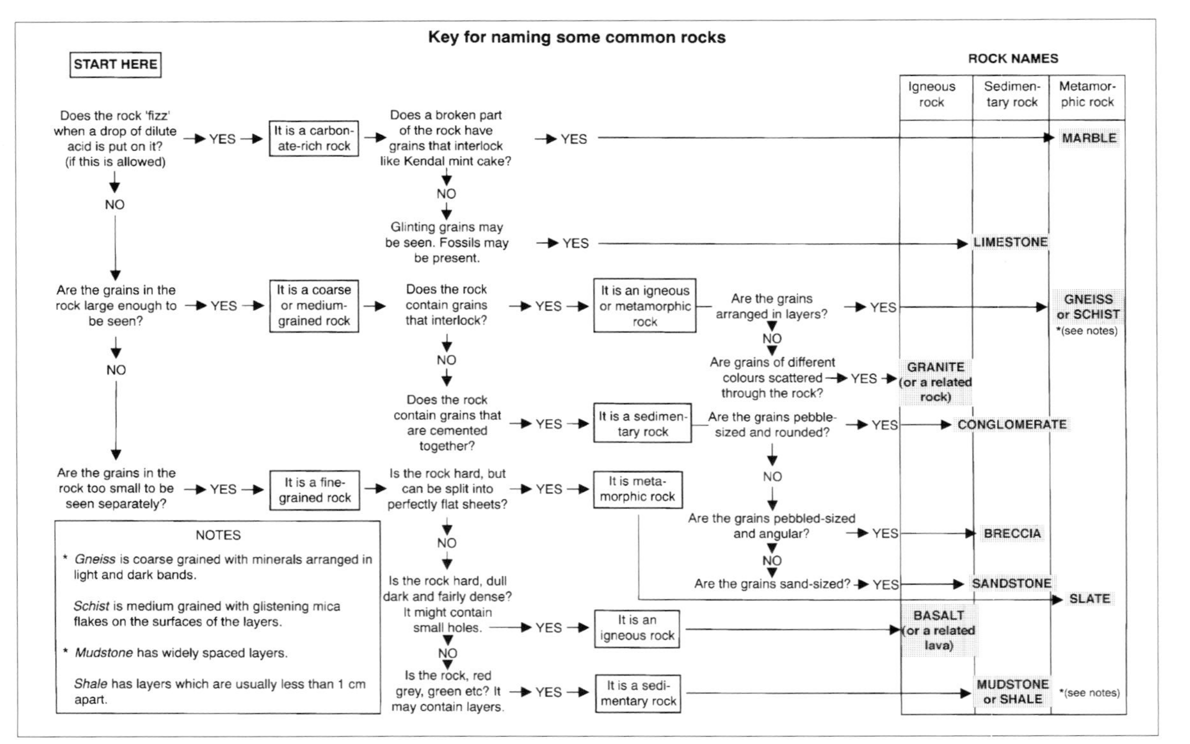

Fig. 16.1. A rock identification key, with the emphasis on the observation of rock characteristics (Modified from *Steps towards the rock face - introducing fieldwork: rocks from the big screen*, published by ESTA 1991).

revealed a list of the most popular project topics, of which 'Materials', 'Homes and Habitats' and 'Local Area Study' present themselves as the most likely topics in which rocks will be studied. A study of the school locality is also a requirement of the Geography National Curriculum at both Key Stages One and Two, to involve fieldwork and nearly all primary school classes will conduct an out of classroom study in the local area during Key Stage One and Two. If teachers are aware of its potential, the local environment therefore provides the 'natural' location to study 'real' rocks.

Secondary schools are traditionally more reluctant to encourage work outside the classroom. The new Science National Curriculum at Key Stages Three and Four, however, refers explicitly to fieldwork as an example of a context for teaching pupils to consider investigative situations where variables cannot be easily controlled, and how data can be collected in such situations. Consequently, secondary schools may feel more obliged to work outside the classroom. Secondary schools, however, have major constraints which prevent the easy organization of fieldwork. A survey of science teachers in 34 urban secondary schools in England and Wales during 1991 identified in rank order the following difficulties in organizing fieldwork: (1) time, (2) staffing, (3) cost, and (4) lack of expertise. These findings suggest that if the study of geology is going to be encouraged beyond the classroom then the traditional environments of rock outcrops, often distant locations, need to be replaced with more realistic venues. The use of the immediate environment of a school is more practical. However, if teaching staff are to make use of this environment, they need advice and help in identifying opportunities which exist in this environment. The survey referred to above revealed that teachers regarded the most useful settings for a local study as, in rank order: (1) housing estates, (2) school grounds, (3) cemeteries and (4) town centres. The ESTA Science of the Earth series of teaching units shows how this can be achieved. For example, units 11 to 14 show how a local rock trail for teaching can be constructed close to a school. It suggests that buildings, roofing materials, path gravel, roadstone chippings, rockeries, paving stones, kerbstones and walls all provide potential geological interest (ESTA 1991). Similarly the units 14 to 16, *How long will my gravestone last?*, give a teaching outline, together with worksheets, for a study of the geology to be found in a local cemetery (ESTA 1988).

Conclusion

The Science National Curriculum requires that geology is taught with a particular focus on the characteristics, properties and textures of rocks. The investigative approach of the Science National Curriculum demands a move away from 'learning by labels' towards developing an understanding of the geological significance of observations through investigation.

One of the main justifications for developing and conserving urban geology lies in its value as a teaching resource for schools, particularly in providing a context in which rocks can be studied in 'real' situations outside the classroom. Such opportunities need to be developed in areas which are easily accessible to schools. In developing resource materials, urban geologists should try to team up with local teachers, who can advise on the most appropriate style and level of material. Urban geology could gain from this a worthwhile purpose, giving teachers a valuable resource with which to deliver what is seen as a problematic part of the National Curriculum, and giving the pupils an enriching, educational experience on their doorstep!

References

Department for Education and Science. 1989. *Science in the National Curriculum*. HMSO, London.

---- 1991. *Science in the National Curriculum*. HMSO, London.

Department for Education. 1995. *Science in the National Curriculum.* HMSO, London.

Earth Science Teachers' Association. 1988. *Science of the Earth 14-16, Unit 1: How long will my gravestone last?* Earth Science Teachers' Association, London.

---- 1991. *Science of the Earth 1-14. Steps towards the rock face-introducing fieldwork: rocks trail*. Earth Science Teachers' Association, London.

17 A version of 'The Wall Game' in Battersea Park

Eric Robinson

Summary

- The educational game known as the 'Wall Game' is described.
- At Battersea Park there is wall made from stone ballast which was dredged from the Port of London during the nineteenth century.
- This wall and others like it provide a superb educational opportunity by providing a variety of rock types in close proximity.

'Geology on your Doorstep' could well take you no further than your local park, especially if it was laid out in Victorian or Edwardian times, when local and municipal pride saw towns in active competition over the novelty and display of their boating lakes, grottoes and rock gardens. Parks were venues used for set-piece pageants, floral displays and promenades. On other occasions the stimulus was to improve the mind, as was the case in the grounds of the new Crystal Palace in Penge, where the principles of geology were demonstrated in company with the discovery of vertebrate palaeontology (Doyle & Robinson 1993). On a smaller scale, other towns had rocky defiles called such things as 'Spion Kop' or 'Khyber Pass' fashioned in local stone or artificial substitutes (Robinson 1994).

Even the containing walls of parks can make a contribution to the need for an urban geological resource for schools. If nothing else, stone walls provide stonework which has stood against wind, rain and frost for perhaps a hundred years and now may begin to show the evidence of physical weathering. In a few exceptional cases, the stone of such walls may provide an even better geological resource.

The potential of stone walls to geology has long been recognized and a simple game has developed known as 'The Wall Game'. It was first played in cathedral precincts where stone from the fabric of the cathedral was often recycled and built into the humbler wall of the surrounding buildings (e.g. Gloucester, Peterborough, Ely and Winchester cathedrals).

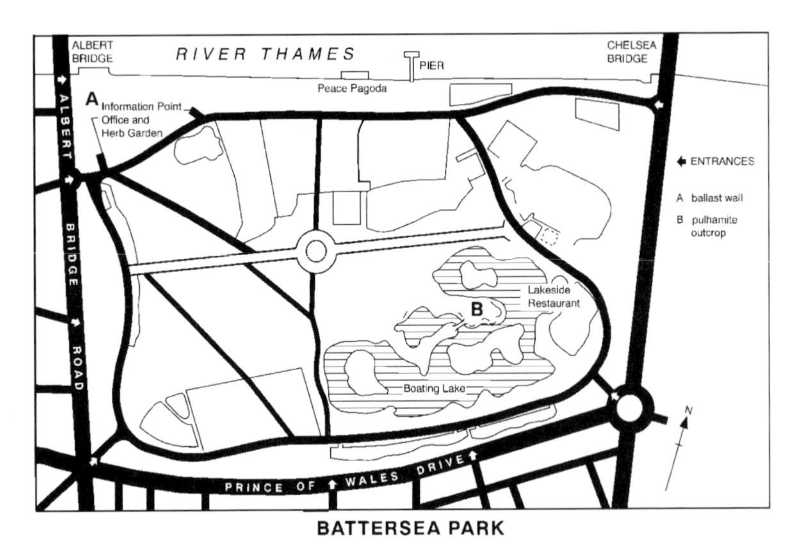

Fig 17.1. Map of Battersea Park.

There are two facets to the game. First, there is the invitation to identify stones which seem different to the eye by colour, texture or durability. Such differences need to be qualified by closer study. Are there shells? Are there sedimentary structures? Can we say whether they are sandstone, limestone, or whatever? The second inquiry is to determine quality of weathering. Is this a good stone or one which has weaknesses? What has caused the break-down? This simple game could be used to teach many of the requirements of the National Curriculum (Hawley 1996). The only requirement is a wall with a range of rock types within it.

In Battersea Park, on the south bank of the Thames opposite the Chelsea Hospital, the chance to play 'The Wall Game' comes in the extreme northwest corner, close to the Superintendent's Office (Fig. 17.1). This is accessible to the public because it coincides with displays of aromatic herbs and shrubs which form a butterfly garden. The same ground is a provision for would-be geologists in that the walls are made up of a variety of stones which immediately catch the eye because of their motley colours and textures. Predominantly grey or white, they are frequently speckled with reds, light browns and a smaller proportion of dark browns. Picking your way through the cars which are parked facing the main north wall, it is possible to study the surfaces at close quarters without interfering with the business of the works yard.

Stone from the seas

Battersea Park was laid out as a park between 1854 and 1856, when the plans of the architect John Gibson were achieved by reclamation of the marshy ground south of the river, by the building of a clay wall and by back-filling with silt and mud excavated from the newly created Surrey Docks down river. Some of the walls within the park were also recycled from material supplied by the Port of London. Stones were obtained from ship's ballast dumped in the middle reaches of the river close to Greenwich by ships before they moved up the Thames to discharge their cargo in the Port of London. Ballast was very important to sea-going sailing ships but less so to steamers which increasingly relied upon water tanks which could be trimmed according to needs. These ballast stones were carried by sailing ships coming into the Thames in the early half of the nineteenth century. It is interesting to think of the trade of that time; a quick survey of the rock types represented within the Battersea wall should allow one to draw some simple conclusions about these trade links.

By far the greatest proportion of the stones are pale-coloured limestones, some of them strongly banded. In most, fossils are not common, but in others, bedding planes are crowded with overlapping shells. Much of this limestone is Purbeck Limestone from the quarries on the Dorset coast near Swanage. A very small precentage is, however, Portland Stone derived from the island quarries south of Weymouth. Yellow limestone in the walls are probably Caen Stone from Normandy, used as building stone for many of the medieval churches in the City and brought from quarries conveniently close to the Channel coast of France. This would have been stone too valuable to be used as ballast, but readily available as older buildings in the City were rebuilt or renovated during Victorian times.

Occasionally, red stones catch the eye, as these contrast to the prevailing paler tones of the wall. Many of these are fire-burnt sandstones which may have been salvaged and recycled from fire-damaged buildings. Intense heat causes changes to the iron cementing minerals, reddening stone which was originally brown or yellow, a change which will be familiar if you look at bonfire sites or open-air barbecue hearths. Reddened stones can show contrasting changes in toughness and durability. Some have literally become fused silica grains; others have lost their cement and have begun to crumble through subsequent weathering. A similar contrast is to be seen in frequent blocks of the grey, gritty sandstone known as Kentish Ragstone, brought in large quantities into London since Roman times from quarries stretching along the length of the North Downs but chiefly concentrated around Maidstone. River boats brought such cargoes down the River Medway into the Thames. Pale-coloured, but flecked with dark green grains of the mineral glauconite, some varieties of Rag are hard and gritty to the touch; others have become soft and crumbly. Both are forms of the Lower Greensand of Kent which can be recognized in the surviving quarries. The harder versions would have been facing stones whilst the softer could have been rubble wall filling.

Two more sandstones are worthy of mention. First, there are a number of stones which may be either well rounded ovate pebbles, or broken, irregular shaped blocks. The stone itself is an amber-orange colour and very compact. It can be quite flinty in its fracture surfaces. These are blocks of quartzite of unknown source, but probably derived from the Thames Terrace gravels of the nearby river meadows of Middlesex. We find them in the gravel pits around Heathrow and they are usually thought of as being derived from the Triassic outcrops of the south Midlands and the north Cotswolds from whence they entered the River Thames during the Ice Age. The second sandstone still to be mentioned is much more a reflection of London streets and is yet another example of early recycling. This is the brown stone which most would recognize as a sandstone and which is broken pieces of paving slabs. Often called 'York Stone', geologically this is Coal Measure sandstone which has been drawn from quarries in the Yorkshire Coalfield since Elizabethan times, and shipped from the Humber as additional cargo on the coal ships coming into the Thames. The stone was first used for gravestones and later for pavings. In the Battersea walls, the flat slab pieces are often introduced into the wall to level up the courses of otherwise oddly shaped stones. A similar role is given to clearly identifiable bricks; either the recognizable yellow London 'stocks' or the dull red facing bricks of a generally higher quality.

Having dealt with the rocks which we could broadly call sedimentary, there remain the dark coloured stones which have a firey origin. First, there are a limited number of blocks of granite visibly consisting of crystals of white feldspar with flecks of black biotite mica and silver flakes of muscovite mica. The granites of Battersea walls are mainly silver-grey granites from Cornwall or Devon, many of the regional quarries being located near the coast behind Penzance and Falmouth. While most of the blocks are sound in character, a few have become deeply weathered and are quite rotten. This makes it easier to appreciate their mineral make-up as mica flakes can be detached with the finger nail, and stumpy crumbs of quartz picked out from the rubble surface.

Turning to the darker coloured stones some have a distinctly rough texture full of closely spaced cavities. This texture is normally referred to as 'vesicular', formed as hot molten lava cools, and the trapped gases within it create a frothy texture. Much the same results occur as metallic slags from furnaces cool and the vesicular lava of Battersea is really modern slags derived from the Thameside foundries of Whitechapel and Wapping. If this seems disappointing, the processes of formation are virtually the same and no one need ever know. As ballast, slags were often used as they form compact shaped cakes which were easier to pack and stow into keel space as well as being dense and heavy for their volume. For this reason, they could be foreign slags rather than purely those of the river.

Reviewing the stone types which we have recognized, it seems clear that the ballast points to a steady trade with ports along the Channel Coast, particularly with Dorset, Devon and Cornwall. There is a contribution, if indirect, from French Channel ports. From the North Sea coast, there is evidence of the coal traffic bringing some quantity of sandstone for paving from the north of England. There is little or no evidence for stone from Scotland or Scandinavia, which, in the nineteenth century were supplying quantities of kerbstone and cobbles for street paving. Perhaps this was too valuable for recycling at so early a time. So, we lack northern gneisses and red granites. The same is also true for the 'bluestones' (diorites) of the Channel Islands of Guernsey and Jersey, and Irish granites.

Equally absent from the Battersea walls are stones such as chalk, slate, ironstone and cementstone 'doggers' (septaria from the London Clay of the estuary), either because they were too poor as ballast, or rejected as walling or building material. In the case of the septaria, a better use was in the manufacture of the earliest quick-setting cements, an industry which was pioneered in the Lower Thames in the eighteenth century. Had the walls been built a generation later, all manner of more exotic stones could have come into the picture. As it is, the range is just sufficient for the beginner-geologist and covers the main classes of rock types other than metamorphic.

'The Wall Game' as a school exercise

Depending upon the size of the class and its aptitude, the simplest project work to set up and complete would be to ask pupils to study a section of a wall, perhaps 2 m wide, and decide from colour, texture and other simple visual criteria, how many different stones they would confidently recognize as 'different'. Ask them to group them into categories as simple as 'sandstone', 'limestone' and possibly 'igneous' (allowing slags to be 'igneous' from a first-principles determination of character). To simplify matters, a cartoon sketch of the stone shapes could be drawn up in advance of the visit (Fig. 17.2). This drawing should have some clearly identifiable

Fig. 17.2. Field sketch of part of the ballast walls on the north side of the works yard.

Fig. 17.3. Photographs of the ballast wall. (Photograph: P. Doyle.)

stone shapes or objects picked out to allow a pupil a fixed point of reference for their observations. The different rock types could be coloured in or shaded with a distinctive pattern chosen by the pupil as representing what they saw. Stemming from their observations and records, it is possible to extend the exercise by trying to match up rock types with their possible sources which could have been trade for coastal shipping in the last century. Localities such as Purbeck, Portland and the granite towns of Cornwall could be searched for on a simple geological map and the necessary connections made. This would extend a knowledge of British geology.

Not all stones are 'sound'. Some are decidedly weakened and crumbling through the agencies of weathering. Weathering may have been increased by the simple fact of placing rocks with different water absorptions alongside one another in this wall. As a result, frost has caused some sandstones to exfoliate and crumble. To a limited extent, pupils can be allowed to finger the softer, crumbling rocks, if only to demonstrate such contrasts. Touching should certainly be encouraged for the slags and vesicular rocks, if only to establish their roughness.

Conclusion

'The Wall Game' is an example of a simple school excercise in an urban environment suitable for teaching some of the geology in the National Curriculum. It is possible in Battersea because of the ballast walls which tell a remarkable story of the commerce of the Port of London. Similar geological resources exist elsewhere: they simply need to be found, exploited and publicized to the local schools. This is perhaps one of the most important contributions which experienced geologists can make to urban geology and to fostering geological interest for future generations.

References

Doyle, P. & Robinson, E. 1993. The Victorian 'Geological Illustrations' of Crystal Palace Park. *Proceedings of the Geologists' Association,* **104**, 181-194.

Hawley, D. 1996. Urban geology and the National Curriculum. *This volume.*

Robinson. E. 1994. The mystery of Pulhamite and an 'outcrop' in Battersea Park. *Proceedings of the Geologists' Association,* **105**, 141-143.

18 Exeter and Norwich: their urban geology compared during medieval, Victorian and Edwardian periods

Jane Dove

Summary

- The geological fabric of Norwich and Exeter, dating from medieval, Victorian and Edwardian periods, are compared.
- Different local stones were used in the two cities during the medieval period, although mostly with the same usage.
- During the Victorian and Edwardian periods, greater transport opportunities led to a greater mix of stone types.
- This survey demonstrates that simple exercises may be developed which could be used to exploit this potential educational resource.

Towns and cities have become increasingly popular venues for exploring geology. A number of geological trails have been devised around urban areas including Stamford (Ireson 1986), Oxford (Jenkins 1988) and Newcastle-under-Lyme (Holloway & Sims 1981). These trails focus on the identification of the rock types, rather than when the stone first appeared in the urban fabric. In contrast, trails devised around central London by Robinson (1984) include this historical dimension. An outline of the general evolution of stone types used in British buildings has been provided by Clifton-Taylor (1987), but the development of the urban fabric is not dealt with specifically. The changes in the urban geology of Exeter is outlined by Dove (1994*a*), but no comparison is made with other urban areas.

This chapter compares the urban geology of Norwich with that of Exeter. These cities were chosen because both have a long history of occupation with buildings constructed in a variety of rock types, but are located in very different geological settings. The objectives of the study are: (1) to identify and account for the similarities and differences in the rock types used in their urban fabrics specifically during the medieval, Victorian, and Edwardian periods; and (2) to offer suggestions as to how the data collected might be used for a student investigation into the factors influencing choice of stone types used in buildings at these times. This approach thereby provides students with information so that they can form their own conclusions, rather than giving them the answers.

Sources and methods

Data on the types of stone used in the urban fabric of both cities, and the location of the visible remains, were collected by field observations and reference to secondary sources. Guides to buildings in both cities provided some useful information (Pevsner 1962; Orbach 1987; Cherry & Pevsner 1989). Information on specific buildings, for example Exeter cathedral (Allan 1991; Laming 1991), was also consulted. *Kelly's Directories* (Kelly's 1910; Kelly's 1937), and journals such as *The Builder*, provided useful information on buildings constructed during the Victorian and Edwardian periods. This information was then used to create a suitable student inquiry task for use in a field based study.

The medieval fabric compared

Local rock types dominate the medieval stone fabric in both cities: flint in Norwich, basalt lava (trap) and Permian breccia in Exeter. Local stone was used for the city walls, medieval bridges, churches, castles, cathedrals and the occasional domestic building. Few domestic buildings survive from before the sixteenth century because most were made of timber and clay. Only wealthy merchants or the clergy could afford homes built in stone.

Flint, the only locally available building stone in Norwich, had to be won from the chalk. It was quarried from pits on the edges of the then city at Kett's Hill, Rosary, Rouen and Earlham Roads (Ayers 1990). The chalk itself was quarried to make lime mortar. In Exeter, volcanic trap was quarried at the site of Rougemont Castle in the centre of the city, and at other volcanic outcrops on the western outskirts. In addition, Permian breccia was quarried within the present suburbs of Exeter at Heavitree and Whipton.

A comparison of the city walls in Norwich and Exeter deserves further mention because both required vast amounts of material and were subject to rebuilding (Fig. 18.1). At Exeter, volcanic 'trap' was used by the

Romans to build the original wall of *c.* AD 200. By the medieval period parts of the wall had collapsed and were rebuilt using Permian breccia then being quarried. However, because vast amounts were required, anything else at hand was used including slates and other material left over from building the cathedral. In Norwich the medieval wall and later repairs were mainly completed using flint. Detailed accounts of the rebuilding of the Cow Tower in the 1390s (Ayers 1990) record the number of cart loads of flint brought to the site. The core of this tower was constructed in flint but faced with medieval brick.

In addition to local material, freestone brought from further afield was used in both cities during the medieval period. Small quantities were used for window and door surrounds in churches and houses in both cities. In Exeter, Beer Stone is common, a type of Cretaceous chalk quarried at Beer, southeast of Exeter. Evidence suggests the stone was brought by sea to Topsham and then carried overland by cart and pack-horse to Exeter. The image screen on the west front of Exeter Cathedral is carved in this stone. Other freestones used for the cathedral included Salcombe Stone, a type of greensand quarried from Salcombe Regis and Dunscombe in East Devon (Figs 18.2 and 18.3). Flint proved unsuitable for dressings and cornerstones for the buildings in Norwich. The freestones used included the Jurassic Barnack, Clipsham and Ancaster stones from the East Midlands, and Caen Stone from Normandy. Caen Stone was brought by sea to Great Yarmouth, then carried up the river

Fig. 18.1. Part of Exeter city wall showing volcanic trap and Permian breccia.

Wensum to the city, and finally transported via a canal dug from Pull's Ferry to the Cathedral Close. In some cases such stone provides clues as to when a structure was built. For example, the triangular headed windows of the round tower of St Mary Colsany appear to be Saxon, but the use of Caen Stone shafts would suggest these were built after the Norman conquest (Atkin 1993). As well as various freestones, Purbeck Marble pillars were transported from Corfe in Dorset for the interiors of both cathedrals.

Both cities still retain medieval streets, Elm Hill in Norwich, Stepcote Hill in Exeter for example. Although subject to repair and replacement, flint was used for the cobbles in Elm Hill, whereas those in Stepcote Hill were composed of a variety of materials including river gravels, city repairs and exotics brought as ballast.

Student inquiry one: the medieval fabric

Initially, the students would be provided with hand specimens of the rock types appearing in the urban fabric of both cities during the medieval period. They would be asked to identify the specimens with the help of

Fig. 18.2. West front of Exeter Cathedral. The image screen is carved in Beer Stone, the rest of the building is largely Salcombe Stone.

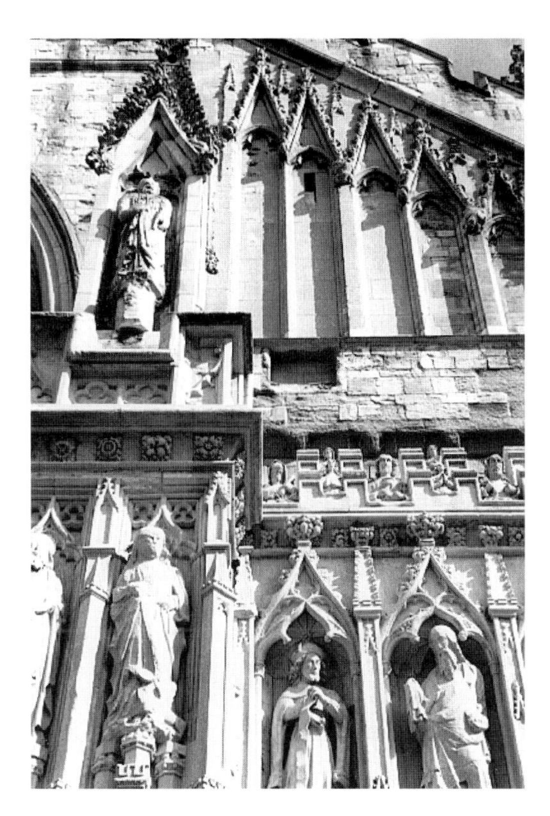

Fig. 18.3. Detail of the west front of Exeter Cathedral. Figures are carved in Beer Stone. Repairs above the screen are in Ketton and Doulting Stone. Differential weathering makes these stand out.

keys like those devised by Harwood (1987). These use grain and crystal size as the criteria for identification rather than colour, a variable which is often deceptive. For example, in this investigation it is likely that the students would not readily identify the volcanic trap as a basalt lava because of its purple coloration. Once identified, the students would be asked to suggest the limitations and advantages of each rock type as a building stone. Evidence based on visual observation, could be supported by reference to books describing the properties of each rock. Flint, for example, being made of silica is very hard and resistant to weathering, but lots of mortar would be needed to hold the nodules together to form a wall. Moreover, although flint could be knapped and used for flushwork, it would not be suitable for carved work. In contrast, the Jurassic limestones would readily form blocks and be suitable for ornamental work. However, its porosity would make it susceptible to physical weathering, and its chemical composition would encourage carbonation and sulphation. Students would be given a simple map of the city with the appropiate buildings marked on it. Ideally, if they lived in one of these cities and were told which buildings to visit, they could conduct their own survey and produce their own map. They would also need to be given information on the available means of transport at the time, and the distance to the quarries. Using this knowledge, the map and the properties of each rock type, the students could then draw their own conclusions as to why specific stone types were chosen.

The Victorian/Edwardian fabric compared

During the Victorian and Edwardian periods, improvements in transport were responsible for introducing a variety of new materials into the urban fabric. A Jurassic freestone, the Bath Oolite, appeared in both cities, and, for example, was used in the Higher Market in Exeter, opened in 1838, and in refacing Norwich Castle between 1833-9. The stone in both cases was brought by water as the railways had not reached these cities. Later, Bath oolite was used for banks in both cities, such as the Crown Bank in Norwich dated 1865, and the alterations to the City Bank of Exeter in 1875 and 1905. Both cities repaired their crumbling cathedrals using Bath oolite and later using East Midland oolites.

In the 1880s work began on the construction of the Roman Catholic cathedral in Norwich. Beer Stone was brought from Devon for its construction although the tower and transepts were built in more resistant Clipsham and Ancaster stones the interior being decorated with pillars of Frosterly Marble from County Durham. In Norwich during the Edwardian period, both Bath and East Midland oolites were used for grand insurance buildings. Monks Park, a type of Bath Oolite, together with Portland Stone, were used in the construction of the Norwich and London Accident Company building, dated 1906, now the offices of British Telecom. The Norwich Union Insurance offices, dated 1901-6, were constructed in Clipsham Stone. Improved transport and the opening of quarries on Dartmoor encouraged granite to appear in Exeter. It was used for pilasters in the Higher Market and later the plinths of monuments. This was not, however, the first appearance of granite in Exeter. The Elizabethan portico, added to the front of the Guildhall was supported on granite pillars believed to have been brought from the eastern edge of Dartmoor by cart (Blaylock

1990). In Norwich Carnmellis, a Cornish granite with massive feldspar phenocrysts, was used for pillars in the Norwich and Accident Assurance Company building already mentioned. The railways introduced Scottish granites into both cities. Pink, Peterhead Granite was used for pillars such as found in front of the United Reformed Church in Norwich, and in Harry Hem's workshop in Exeter. In Norwich pillars of grey, Scottish granite were constructed at the entrance to the London and Provincial Bank dated 1907, and now a branch of Waterstone's.

More varied sedimentary rocks were also used. For example, Kentish Ragstone, a Cretaceous sandstone, was occasionally used in Norwich in the construction of St Matthew's Church dated 1851, and now converted to offices. Westleigh Stone, a Carboniferous limestone quarried near Wellington, was used in the construction of several buildings in Exeter from the 1850s including St Luke's Training College, St David's railway station, and Wonford Hospital. It was also used in the construction of several new churches around the edge of the city centre such as St Michael's in Dinian Road. Though dressings were often in Bath oolite, Devonian limestone quarried near Torquay was used to face Emmanuel Church, and a warehouse on the quay.

Local stone continued to be used in both cities during the Victorian period, although more so in Exeter. In Norwich, flint was used in the construction of several churches around the edges of the city centre, including St Mark's dated 1844, and Christ Church dated 1873. Yellow brick was used for dressings for St Mark's, and Bath Oolite for Christ Church. In Exeter, volcanic trap and Permian breccia continued to be used for buildings such as almshouses and a few churches. The Royal Albert Memorial Museum was mainly constructed in volcanic trap with dressings in Devonian Limestone from Chudleigh and Bath Oolite.

Brick became more widespread as a building material in the nineteenth century in both cities. It was used for Victorian terraced housing, schools, chapels, hospitals, factories and some public buildings. Light yellow bricks called 'cossey whites' became popular in Norwich in the 1830s, used for example in the United Reformed Church. Red brick was used for factories in Norwich such as Jarrold's Silk Mill dated 1839, and in Bullards Brewery dated 1868. Other notable buildings constructed in red brick included the Agricultural Hall (now Anglia Television) dated 1882, the Norfolk and Norwich Hospital dated 1879, Thorpe railway station rebuilt in 1886; and in the Norwich Technical Institute constructed in

1899. In Exeter, brick was popular for schools such as Maynard's, hospitals such as the Exe Vale, and breweries such as St Anne's. In both cities, these red brick buildings were often dressed with Bath oolite or white brick. Terracotta, popular in the late Victorian period also appeared in both cities. In Norwich, the architect Skipper used red terracotta to decorate his offices dated 1896, while in Exeter, Wilkinson used yellow terracotta for the Exe Valley Hospital.

Student inquiry two: the Victorian/Edwardian heritage

In this activity students would be encouraged to use secondary data sources to compare and contrast the nature of the Victorian and Edwardian geological fabric in both cities. Directories of each city (Kelly's 1910, 1937) provide a list of the main buildings constructed during this period and some reference to construction materials. Occasional articles in the journal *The Builder* describe buildings constructed or altered in both cities during this time, although those chosen tend to be the largest or the most architecturally interesting. Local newspapers, for example the *Devon and Exeter Gazette*, would also provide useful material. In using these data, the student would need to take into consideration the fact that some buildings would have since disappeared: the High Street in Exeter was heavily bombed during the Second World War removing buildings such as the General Post Office, complete with columns of Aberdeen Granite. Ideally, the students should look at these buildings in the field, identifying the stone types with the aid of a key. The activity would, however, not be dependent on this; instead students could be given hand specimens of the rock types used by the Victorian/Edwardian builders to identify. Seeing specimens would be important, because students hold their own perceptions about rock types. For example, the massive, grey, crystalline appearance of the Devonian and Carboniferous limestones seen in Exeter would probably not fit the student's perception of a limestone (Dove 1994*b*). The data collected from the secondary sources could be summarized into a table. Using this, and another table drawn up for Norwich, students could analyse the similarities and differences in the building materials used in the cities at this time and suggest reasons for the patterns they identified. They could also investigate whether there was an association between material used and building type. In approaching this, they would need to consider when the buildings were constructed because different materials became available and fashionable at different times. The students could also plot the distribution of the buildings they had identified on a map. Clearly not all buildings constructed between 1837 and 1911 would be shown, but it would enable the students to see how the urban geology reflected the spread of the city beyond the medieval core at this time.

Conclusion

Although both cities used different stone types for their medieval fabrics, the principles governing their use appeared very similar. The cost of transporting a heavy material such as stone confined early buildings to using local rock types. Even the cathedral, the most important building in each city, used large quantities of local stone. Moreover, the concentration and coloration of the local stone, black flint nodules encrusted with white in Norwich, purple trap and red sandstone in Exeter, still impart a very visual image of the location of the medieval core today, despite the fact that local materials have been used for occasional buildings in the suburbs of each city. The Victorian and Edwardian periods saw both similarities and differences in the new materials used in each city, common materials being Jurassic oolites, with East Midland varieties being more associated with Norwich. Although brick became more widespread in both cities, Exeter used more stone than Norwich for some of its larger buildings, a reflection of accessibility to local sources. The patterns which result provide good case studies on which to base student inquiries about the availability of stone in different periods.

References

Allan, J. 1991. A Note on the Building Stones of the Cathedral. *Medieval Art and Architecture at Exeter Cathedral.* British Archaeological Association, Conference, **11**, 101-108.

Atkin, M. 1993 . *Norwich history and guide.* Sutton, Stroud.

Ayers, B. S. 1990. Building a Fine City: the provision of flint, mortar and freestone in medieval Norwich. *In:* Parsons, D. (ed.) *Stone quarrying and building AD 43- 1525.* Phillimore, Chichester, 217-227.

Blaylock, S. R. 1990. Exeter Guildhall. *Devon Archaeological Society Proceedings,* **48**, 123-178.

Cherry, B. & Pevsner, N. 1989. *The buildings of England: Devon.* Penguin, London.

Clifton-Taylor, E. 1987. *The Pattern of English Building.* Faber and Faber, London.

Dove, J. E. 1994*a*. *Exeter in Stone* . Thematic Trails, Oxford.

---- 1994*b*. *Student identification of rock types*. Unpublished manuscript, School of Education, University of Exeter.

Harwood, D. 1987. Rock identification: an appropriate skill for ten and eleven year olds? *Teaching Earth Sciences,* **12,** 58-67.

Holloway, S. & Sims, K. 1981. *A Guide to the building stones of Newcastle-under-Lyme.* Keele University Library, Occasional Publication, **17**.

Ireson, A. S. 1986. *The Stones of Stamford.* Stamford Development committee, Stamford.

Jenkins, P. 1988. *Geology and buildings of Oxford.* Thematic Trails, Oxford.

Kelly's. 1910. *Kelly's directory of Devonshire.* Kelly's, London.

---- 1937. *Kelly's directory of the counties of Cambridge, Norfolk and Suffolk.* Kelly's, London.

Laming, D. J. C. 1991. The building stone and its quarry. *In:* Swanton, M. (ed.) *Exeter Cathedral: a celebration.* Dean and Chapter of Exeter, 65-73.

Orbach, J. 1987. *Victorian architecture in Briain.* A and C Black, London.

Pevsner, N. 1962. *Norfolk, North East and Norwich.* Penguin, London.

Robinson, E. 1984. *London illustrated geological walks book 1.* Scottish Academic Press, Edinburgh.

19 Earth science sites in urban areas: the lessons from wildlife conservation

George M. A. Barker

Summary

- Earth heritage conservation has long lagged behind developments in wildlife conservation.
- Some salutary lessons for the promotion of earth heritage conservation may be learnt from past attempts to promote wildlife conservation in urban areas.
- The key to success has been found to be local community involvement. If a site is valued and becomes part of the community, then it will be defended and conserved as part of that community.

Although important wildlife sites are found in urban areas, those which have been notified as Sites of Special Scientific Interest (SSSI) are relatively few and far between (Table 19.1). These are usually relics of former rural habitats around which urban development has swept leaving them as islands in a sea of buildings. For these few sites the nature conservation legislation gives clear protection.

The majority of the habitats on which wildlife in urban areas depends do not enjoy this protection and until quite recently most naturalists and those concerned with nature conservation, paid them little attention. They

Table 19.1. Sites of Special Scientific Interest in England 1990.

Total	Urban*	Urban fringe**	Urban coastal	Urban fringe coastal
3 383	101	755	21	95
743 141 ha	6 094 ha	170 993 ha	3 702 ha	101 236 ha
% by number	3	22.3	0.62	2.8
% by area	0.82	23	0.5	13.6

*Urban is defined as sites embedded in built-up areas > 1 km^2. **Urban fringe is defined as sites within 1 km of an urban area.

include private gardens, public parks, land alongside transport corridors, canals, industrial pools, demolition sites, old mines, quarries, industrial waste, and many other places including the buildings themselves. Their importance as part of the national resource of wildlife is still not adequately recognized. Work, however, by professional scientists and amateur naturalists over the past 30 years, and notably during the late 1970s and early 1980s, has begun to show that many of these sites have at least high importance locally for wildlife and that nationally uncommon species live in some (e.g. Teagle 1978; Sukopp & Werner 1982; Barker 1987). Attempts have therefore been made to give these sites and locations some level of protection against development and damage to their interest (e.g. Box et *al.* 1994).

In an attempt to provide this protection two problems were met immediately. First, the Nature Conservancy Council was not prepared to devalue the SSSI currency, as it saw it, by including such sites among those formally notified. It was, however, prepared to support the idea of a 'second tier', in practice often extending to a 'third tier', of sites of local importance for nature conservation so long as this did not add to the range and number of sites covered formally by the nature conservation legislation. Secondly, although local authorities have a general duty to consider nature conservation when carrying out their functions, they certainly did not see themselves as nature conservation organizations. They would play some part, but depended on the non-governmental and governmental nature conservation organizations to take the lead and to advise them what they were required to do under the terms of the nature conservation legislation. For example, the hope that sites not notified as SSSI would be protected by local plan policies often was not fulfilled.

Introduction of societal values

Most local plans now recognize, define and give protection to a considerable number of sites of local importance for wildlife and of green networks in urban areas. This shift in policy is the result of complex influences. High among these, however, is an acceptance that such sites are important to people as well as to wildlife, and that their importance to people depends in no small part on the wildlife they hold. This shift is important because it means that the local authorities, which are recognized as key organizations here, are not acting primarily as nature conservation organizations in protecting and enhancing these sites, but as servants of the communities whose welfare they are in business to look after. Coupled with this is an increasing acceptance that daily contact with nature is beneficial to human physical and mental well-being despite the fact that the research base for this belief is still incomplete (Harrison *et al.* 1987; Rohde & Kendle 1994).

There can be little doubt that the research about human well-being, the additional landscape and amenity values of natural open spaces in urban environments, their possible roles as pollution sinks, their educational values and so on have influenced thinking. So too has the fact that the statutory nature conservation agencies have made it clear that they feel the social values of contact with wildlife are a legitimate and important part of nature conservation.

At least as important, perhaps more so initially, was the rise of the urban wildlife groups as organizations prepared to make nature conservation a local political issue (Smyth 1987). When an articulate and influential group of local voters comes onto the scene, local politicians are well advised to take note of their concerns. The urban wildlife groups began as elite organizations and some remain so, but others have become or have generated genuine grass-roots support, and local communities have become directly concerned in protecting, managing and using sites.

Present planning guidance

All of this has led to a change in the content of statutory local plans. In the 1960s and 1970s mention of nature conservation was simply confined to a routine reference to SSSI. There are now often extensive sections with policies on nature conservation. These often relate to relatively large areas of open land, to the enhancement of this land in the course of development or redevelopment and to the association of human and economic well-being with environmental conservation. These policies are often found in free-standing and detailed nature conservation strategies published, as non-statutory planning documents, by an increasing number of urban local authorities. These strategies are based on detailed surveys of the natural and other open habitats in the district concerned (Barker 1984; Salmon Widman & Associates 1994). These surveys not only underpin these statements, but also provide a wide range of local and national nature conservation organizations with the opportunity to contribute knowledge, data, ideas and practical support.

In England these ideas have been taken by the Department of the Environment (1994) and published in their *Planning Policy Guidance - Nature conservation (PPG 9)* giving the local authorities a strong focus and incentive for including policies to protect locally important sites, corridors and green networks in their statutory plans. It also opens up the possibilities for forward planning. In this connection it is worth mentioning the discussion paper of United Kingdom: Man and the Biosphere Committee's Urban Forum on the standards for the provision of natural green spaces in urban areas (Box & Harrison 1993). This argues that if contact with nature is important for people in the course of their daily lives, then accessible natural green spaces must be provided in urban areas. It suggests that at least two hectares of accessible natural green space be ensured within 0.5 km of every home and that there should be one hectare of statutory Local Nature Reserve for every 1000 people. In addition, it suggests targets of at least one 20 ha site within 2 km, at least one 100 ha site within 5 km and at least one 500 ha site within 10 km of all residents. Clearly earth science conservation can be built into programmes like these.

Positive approaches

The current situation is that protection is given where none used to be. Traditionally, the perceived goal has been to make sure that sites are not built on or developed in any other way. Although some management is often seen to be needed, this is an issue secondary to securing the site itself. The idea of actually using development as a tool in wildlife conservation is a recent one, but in the present political and economic climate it is an important step to take. There is a wide range of options. We are all familiar with the possibilities of seeing some mineral extraction sites restored to nature conservation end-uses and this has become a common condition of planning consent. The reclamation of derelict land for nature conservation, something which nature has been doing all by itself for years, has been recognized by the Department of the Environment as the most cost-effective approach in many circumstances (Land

Capability Consultants 1989). The use of planning gain, that is agreements between the local authority and a developer under Section 106 of the Town and Country Planning Act 1990, is becoming more usual. This effectively involves a trade-off in which development is permitted under certain conditions which may include making land over to the local authority or providing some agreed facility for wildlife. The use of conditions favouring nature conservation in giving planning consent is endorsed by the Department of the Environment in PPG 9. The way is wide open for local authorities to insist on seeing green networks enhanced and extended, new habitats for wildlife created and environmentally friendly approaches taken as a part of redevelopment or new development. A problem remains, however, because developers and local authorities have difficulty getting clear and precise advice. This reflects the fact that our understanding of the operation of green networks (e.g. Dawson 1994), the natural make-up of urban habitats (e.g. Gilbert 1989), the rationale behind habitat creation and similar science based questions remain unclear. This problem is compounded by those concerned about nature conservation seeing developers as the enemy and, it must be said, *vice versa*. There is, however, a strong element in nature conservation which now takes the view that the knowledge and skills of ecologists must be brought into play to help developers and land managers achieve their goals more efficiently and with greater benefit to the environment than they would be able to without such a liaison. This is in sympathy with current political thinking which prefers policies for site protection to begin 'Development will normally be permitted unless...' to 'Development will not normally be permitted...'.

Concepts of community

One of the most important concerns of urban wildlife conservation is the values of natural green spaces to society. They have proved important in developing ideas and action on the biological aspects of nature conservation and will prove so too, indeed have proved so, for earth science aspects. In addressing societal values, the first question to clarify is: what is the community whose values we are examining? Experience in wildlife conservation suggests we are dealing with two kinds of community: (1) *communities of interest*, including the natural history societies, wildlife trusts, urban wildlife groups and increasingly other associations such as broad-based countryside amenity and environmental groups; and (2) *communities of place*, the people, whatever their interests, who inhabit a particular location.

There is, of course, overlapping membership of these communities. Within each basic type of community there are variations in scale. A natural history society, for example, can be a global, national, regional or local one. A community of place can encompass all humanity, a neighbourhood, a street or a part of a street.

It is with communities of interest that wildlife conservation has felt most comfortable. Urban habitats have, however, been consistently undervalued by this community. The urban wildlife groups of the 1980s began to change this and broaden the field of interest to include their perception of societal values. At best these values were in harmony with those of communities of place, but in many instances were far removed from the perceived priorities of local inhabitants. It was soon realized that if the people whose daily lives impinged upon a site were not supportive, then the best laid plans for site conservation were at risk. Schemes imposed from outside often fail where ones growing out of the local community succeed. In Northamptonshire the County Council's proposals for a network of picnic sites was greeted with hostility by landowners and suspicion by local people, while at the same time some communities were asking for advice and money to set up pocket parks. Here the local authority had the good sense to scrap its own scheme and to develop the local initiative, and recently the county's fiftieth pocket park was set up (Anon 1995). In the days of the Manpower Services Commission flying squad teams, school nature areas sprang up overnight and were as rapidly demolished by the pupils. Whilst those planned and developed by pupils, parents and staff endure. There needs to be some sense of ownership among the community of place.

Community ownership and control

Communities of place have very complex agendas. It is rare for any site to be able to meet all the goals a community has set. Provided that the community accepts the place as part of its territory and that those who obtain most from a site are accepted as members of the community, there is minimal conflict and even support from those who make little or no use of the site. For example, the regulars in the village pub raising money for restoration of the church when few of them use it. Used or not, the church is part of the village fabric and its congregation a part of the community deserving support from everyone.

It is important in setting up a nature reserve to find the threads which link it to local people and to weave these into strong ties. In wildlife conservation the following have been identified as important: an element of control through public consultation and volunteer activity; explanation, people value things more if they understand them; identification of community priorities and needs and use of the site to meet these needs, if it is possible, without significant damage to the site's interest; connections with the community's past; a clear focus for suggestions, comments and criticisms; using the site, if ony occasionally, as a focus for community activities; job-creation and income generation; good publicity especially in local newspapers; absence of an exclusive controlling clique; and a feeling of safety when on the site.

A large urban Local Nature Reserve (LNR) is likely to be successful therefore if it: (1) has wardens drawn from local people and accepted as part of the community; (2) is used by schools, ramblers and for local social events; (3) preserves traces of bygone farming or industrial activity; (4) has good explanatory leaflets; (5) has a friends or user's group; and (6) has wardens who make sure events, such as sightings of unusual animals and visits by interesting people, get into the papers. If this pattern is followed then anything perceived as a threat to it or its staff is seen as a threat to the whole community and is roundly opposed. Saltwells LNR in Dudley provides a good example, where the suggestion that part of the site may become an open cast coal mine brought vigorous opposition from local people as well as from the communities of interest (Smyth 1990).

The example of Wybunbury Moss National Nature Reserve in Cheshire is informative too. In the early 1970s the local community which borders this small floating bog on three sides was so alienated by the conservationists that the Parish Council wrote an open letter to the Press complaining about the site. The Council did not know how otherwise to contact the Nature Conservancy Council (NCC) since it employed no local staff. Strenuous efforts were then made by the NCC to explain the reserve to local people. There were public meetings, open days and the village school became involved and helped plan the site management. NCC staff became accepted as part of the village community. When, in the 1980s, it was found that septic tank overflows threatened the Reserve, it was the Parish Council which made robust approaches to the District and County Councils, the Northwest Water Authority and the local MP which helped bring about a solution to the problem which the NCC alone would have found hard to achieve. A threat to the site had become a threat to the village. It was, by then, 'our reserve' to local people.

Wardens

Although many sites are seen by local residents as part of their home territory, it is the exception, rather than the rule, that the land is owned by the local community. Northamptonshire's Pocket Parks are among these exceptions, being owned by community trusts and managed by larger 'friends' groups. The land may be owned or leased by a community of interest, such as a wildlife trust, by private owners who permit access, or, particularly in urban areas, by the local authority. In all these cases there are often individuals who act voluntarily or who are paid to warden the site. They are very important.

In urban situations they can be on-site teachers and sources of information as well as policing the place. Their constant presence turns sites, seen previously as unsafe, into havens where children and adults who feel vulnerable can be secure. By working to a policy of talking to everyone they meet, they perform a valuable social function, become recipients of information as well as donors of it, and are soon familiar members of the community whether they actually live in it or not. Particularly where they are local authority employees they can form a close link between the community and the authority and become important in community development.

Unfortunately their wider potential is often unrecognized. This means that they are evaluated simply as site staff maintaining a property and, as such, are expensive and expendable when budgets are under pressure.

The broader and longer term picture is an important one. Wardens give the potential for the local authority to become closer to the community. This transforms the sites from simple amenities into the focus for community development and improved local authority efficiency. It puts a new and high value onto such places and their wardens.

All this depends on the staff employed or the volunteers concerned. There is the risk of a 'club' forming which, whether by intent or not, discourages those outside it from becoming involved in the site. There is, too, the risk that the wrong person in the post can alienate the local community rather than win its support. These risks are worth taking in view of the enormous benefits a good team of wardens can bring. It is very hard without them to make nature conservation a community issue.

What can earth science learn from this?

Wildlife conservation has many parallels with earth science conservation. It is reasonable to suppose that the lessons learnt in one branch of nature conservation apply to another. The most important lesson learnt recently in wildlife conservation is that nature conservation has much more to learn from social, psychological and behavioural sciences. Sites seen as locally significant by communities of interest, depend for their safeguard on the values and uses given by the local communities of place. The interest and support we need to ensure that the natural values of any site are respected can rarely be generated simply by explaining these values to local residents in the hope that they will reflect the enthusiasm specialists have. A broader-based approach is needed in which the significance of the site to local people is explored together with the potential for its use in addressing community needs.

Earth science sites need therefore to accrete around them a suite of other interests. Among these are landscape, local amenity, wildlife, industrial archaeology, site history and recreation potential. Since almost all of these depend to some extent on the geology, earth sciences have a good basis on which to gain community interest. Anything which can be connected with community history is very valuable and many geological sites have clear links. Among the wide variety of special interests which any site can attract, it is likely that one will take the lead in trying to secure and manage the site. Which takes the lead is relatively unimportant provided it does not see itself as exclusive. This applies just as much to nature conservation as it does, for example, to recreation.

The site needs to be seen as safe and welcoming. It must also provide stimulus, mental as well as physical. There is a myth that people visit natural green spaces for relaxation. All the research suggests that they go to be stimulated: they go for the adventure (e.g. Hull & Harvey 1989).

Somebody needs to act as the link between owners, managers, specialist interests and the local community. They need to be accessible and skilful, capable of motivating others without excluding anyone. Local people need to be informed about all aspects of the site and invited to take an interest and contribute information. The local press has a part to play as well as leaflets and information boards. The local school is often a way of getting information into local communities.

This emphasis on communities of place should not disguise the importance of communities of interest. The officials concerned with earth science conservation and with wildlife conservation have worked so long with interest groups as to be very familiar with their importance. It is on them that reliance must be placed at local level for a good deal of specialist advice. The professionals here have a job to make sure their allies have access to the very best and most recent information and thought in their subject.

More difficult in the context of communities of interest is influencing and motivating those in different interest or professional groups - planners, industrialists, developers, farmers and foresters - to take nature conservation interests into account. Both wildlife conservation and earth science conservation recognize the same key groups and in particular local authorities, Government departments and the owners of the sites concerned. The significant change in addressing these audiences has been to include the benefits nature conservation brings to society in our definition of nature conservation as well as the maintenance of natural heritage for its own sake. While in some respects earth science conservation got there ahead of wildlife conservation in emphasizing its importance in the production line of the geologists needing to track down mineral resources, in others it has perhaps lagged behind, notably in appealing to local communities of place. Engaging the local community is to tap a powerful stream of local and national political influence.

So far as local authorities are concerned it is helpful to show the relevance of nature conservation to people, to link aspects of it to industries like tourism, to emphasize its positive contribution to new development and pollution control, and to bring out its educational potential. The systematic survey of the resource, however, was seen as vital because it gave backing to protective policies and confidence in defending chosen sites at public inquiry. Linked to this was the evaluation of the resource.

Here wildlife conservation made mistakes and, in particular, in agreeing to draw up hierarchical systems. Experience here suggests it is counterproductive to nature conservation, however attractive to planners, to produce hierarchical lists. For this reason in wildlife conservation the preference is now to list SSSI and then sites of importance for nature conservation in the locality concerned. These non-statutory sites are as important as the SSSI but for different and complementary reasons. They are *not* second grade sites.

Finally, wildlife conservation has made the connection between personal action and the feel-good factor. It is satisfying for an individual to do something, or not do something, which is recognized to be good for nature conservation. It is especially satisfying if this is praised by people who are respected in this field. For an individual to contribute in a way they feel happy about often comes down to something outside sites of recognized importance. The wildlife gardening movement of the 1980s exploited this and was perhaps slow in realizing that influential individuals can be recruited in this way. I am not sure what the equivalent in earth sciences is to wildlife gardening but it will be useful if found. I would encourage interest in earth sciences away from the special sites. In exploiting cemeteries and building stones, geologists have the right idea. They could perhaps make better links with human and community interests than they do. It is interesting that marble, for example, was brought to a particular place from, say, India, but why for *that* particular gravestone? Who first imported it? Who made it fashionable?

In this case, however, earth sciencists can probably teach wildlife scientists, judging by a lot of recent nature trail literature. It is so frustrating to meet the bald statement 'the banks of the river are covered with Himalayan balsam' when so many fascinating stories can be added to engage the interest instead of giving rise, in me at least, to the seditious thought 'So what?'. Nature conservation cannot afford to do this either in connection with wildlife or earth sciences when we depend on the reader for support.

It is, unfortunately still a minority of practitioners who have discovered that the right messages are that nature conservation is *interesting* and is *fun* where the uncommitted are concerned. Saving the world is strictly for the committed!

References

Anon. 1995. Pockets full of countryside in Northamptonshire. *Urban wildlife news,* **12,** in press.

Barker, G. M. A. 1984. Nature conservation abroad. *The Planner,* **70**, 21.

---- 1987. European Approaches to Urban Wildlife Programs. *In:* Adams, L. W. & Leedy, D. L. (eds) *Integrating man and nature in the metropolitan environment. Proceedings of a National Symposium on Urban Wildlife. 4-7 Nov. 1986, Chevy Chase, Maryland.* National Institute for Urban Wildlife, 183-190.

Box, J., Douse, A. & Kohler, T. 1994. Non-statutory Sites of Importance for Nature Conservation in the West Midlands. *Journal of Environmental Planning and Management,* **37**, 361-367.

---- & Harrison, C. 1993. Natural spaces in urban places. *Town & Country Planning,* **62**, 231-234.

Dawson, D. 1994. *Are habitat corridors conduits for plants and animals in a fragmented landscape? A review of the scientific evidence.* English Nature Research, Report **94**, Peterborough.

Department of the Environment. 1994. *Planning Policy Guidance: Nature Conservation.* PPG 9. HMSO, London.

Gilbert, O. L. 1989. *The Ecology of Urban Habitats.* Chapman and Hall, London.

Harrison, C., Limb, M. & Burgess, J. 1987. Nature in the City - Popular Values for a Living World. *Journal of Environmental Management,* **25**, 347-362.

Hull, M. & Harvey, M. 1989. Explaining the emotion people experience in suburban parks. *Environment and Behaviour*, **21**, 323-345.

Land Capability Consultants. 1989. *Cost effective management of reclaimed sites.* HMSO, London.

Rohde, C. L. E. & Kendle, A. D. 1994. *Human well-being, natural landscapes and wildlife in urban areas. A review.* English Nature, Science **22**, English Nature, Peterborough.

Salmon Widman & Associates. 1994. *Nature Conservation Strategies: the way forward.* English Nature, Peterborough.

Smyth, B. 1987. *City Wildspace.* Hilary Shipman.

---- 1990. *The Blackbrook Valley Project. 1981-1988.* Urban Wildlife Now **6**, Nature Conservancy Council, Peterborough.

Sukopp, H. & Werner, P. 1982. *Nature in cities.* Nature and Environment Series **28**, Council of Europe, Strasbourg.

Teagle, W. G. 1978. *The Endless Village: The wildlife of Birmingham, Dudley, Sandwell, Walsall and Wolverhampton.* Nature Conservancy Council, Shrewsbury.

20 Shifting the focus: a framework for community participation in earth heritage conservation in urban areas

Greg Carson & Mike Harley

Summary

- To be effective, earth heritage conservation needs to be focused on people: a people-centred strategy.
- Raising public awareness of geology is essential to its effective conservation; the urban environment is the best place to do this since it is here that most people live.
- Active local participation in conservation is considered to be essential for it to succeed.
- The range of organizations and groups which provide an opportunity for this participation is reviewed.

The conservation of our earth heritage forms a vital part in the conservation of the natural environment. Until recently, however, effort has been focused almost entirely on conserving sites for their intrinsic scientific value. Such sites may be designated as Sites of Special Scientific Interest (SSSI) and given statutory protection within the planning system. Equally, many non-statutory sites important for earth heritage conservation are designated as Regionally Important Geological/Geomorphological Sites (RIGS) on purely scientific merit, although RIGS may also be designated on the basis of their aesthetic, educational, cultural and historical value.

Considerable resources have been, and are being, used to protect sites that are of value only to scientific specialists in a particular field. This community of scientists forms only a small percentage of our total population, and of these, even fewer are directly involved in earth heritage conservation. Resources may be better directed towards: (1) seeking to involve those with an existing interest in nature conservation in earth heritage conservation; and (2) finding imaginative ways to raise the awareness of earth heritage conservation with a much larger proportion of the population. This is especially relevant in urban areas where 80%

of the population of Britain dwell. In short, to concentrate resources into a people-centred conservation strategy.

A people-centred conservation strategy

A people-centred strategy may be approached on a variety of scales. At a large-scale one can use the media to capitalize on 'box-office' successes, such as *Jurassic Park* and *The Flintstones*, in order to raise awareness of geology. For geological conservation to be truly effective, however, we need to encourage *local* people to be aware of this aspect of their *local* environment. This has been the basis of the RIGS movement for a number of years and in England alone, over 1700 locally important sites have been formally notified to local authorities by local groups. Although the majority of these sites are designated on the basis of their scientific and educational value, the spectrum of criteria for site selection is broad, ranging from those sites of purely academic interest, to those which are of purely aesthetic value. In terms of conservation, the qualities of our geological heritage that lie outside the scientific realm is one of the most effective ways in connecting local citizens to geological conservation.

Traditionally, earth heritage conservation has been effected by the *specialist,* either in a professional or voluntary capacity, as a knowledgeable member of a local wildlife trust, geological society, a museum curator or local amateur. To achieve the full potential of earth heritage conservation, however, attention has to be focused on the remainder of the population. This is essential for sustainable conservation of the geological resource.

Much of the effort in earth heritage conservation has been concentrated in the countryside. Towns and cities, however, contain the majority of Great Britain's population (80%) and it is here that a people-centred approach to earth heritage conservation may have its greatest impact. Within such areas, it is social and industrial development that tends to mask the local geology and reduce the connection people have with their natural environment. Increasing awareness, empowering and involving those *locally* is one way of enabling people to gain a sense of place within their immediate environment, and from this to derive a respect for their natural and historical heritage.

Within urban areas, there are organizations in place for encouraging participation in and raising awareness of environmental and conservation matters. These organizations cater for different levels of interest, from those dedicated to urban geology, to those which can help to raise awareness of earth heritage conservation by means that are virtually 'subliminal'. Traditionally, it has been the former which have been most involved in earth heritage conservation, and will continue to play an active role in the future. The full potential, however, of earth heritage conservation will only be realized if interactive partnerships between these various groups are established. In the remaining part of this paper these groups are discussed.

Urban geological societies

There are, in England, more than 25 local geological societies based in urban areas, some of which have been active in urban earth heritage conservation for a number of years. Of these, the Black Country Geological Society has been a highly effective force on the ground (Cutler 1994, 1996). In Dudley the Society has been involved in a wide range of activities, including site enhancement and site interpretation. Moreover, the Society has linked the industrial and geological heritage of the Black Country with its local history. This has been instrumental in raising the profile of the geological resource within the local community to the extent that it is now an integral part of the local culture.

In addition to the conservation of specific sites, urban geological societies have been active in broadening public understanding of the way that geology interacts with everyday life. Promoting awareness of building stones and the variety of rock types found in churchyards is one way in which local geological societies enable local people to gain a greater appreciation of their local area.

The British Trust for Conservation Volunteers

The British Trust for Conservation Volunteers (BTCV) provides hands-on experience of nature conservation and land management in rural and urban areas. Such experience is important in urban areas where it allows people to interact, on a practical level, with their local environment. The BTCV has been involved in a range of geological conservation projects, for the most part in rural areas. The BTCV has, however, also carried out some clearance work in urban areas, such as at Pennyquick Bridge in Bath, revealing a Lower Lias section, and at Cherry Gardens Railway Cutting in Bristol, exposing Pennant Sandstone, faulted against Triassic red marls, and overlying Rhaetic/Liassic limestones and clays. By involving the BTCV with geological conservation, awareness of its importance is spread to a wide range of people with little or no prior knowledge of geology.

Common Ground

What about those who have no interest in their local environment? How do we inspire awareness in the wider public audience? Linking an appreciation of landscape with earth heritage conservation is being taken forward by organizations such as Scottish Natural Heritage in their booklet *Edinburgh - a landscape fashioned by geology* (McAdam 1994). This approach, however, can only be adopted where towns and cities have striking relief which can be coupled to the geology. Many urban areas possess monotonous geological and geomorphological conditions. Under such circumstances, using building stones and churchyards to highlight the importance of our geological heritage is one way forward. This approach, however, can only work where the person is an interested information-seeker (Keene 1994): somebody who will buy the book or attend the urban excursion. To really broaden earth heritage conservation, we have to explore other methods of allowing people to interact with their environment. Here we have much to learn from Common Ground.

Common Ground works through a variety of different projects which encourage local people to become aware of and value distinctive features in their everyday landscape. They are active mostly in rural areas on a parish level and use art and sculpture, frequently with a geological angle, to enable people to rediscover their sense of place within the environment. For example, in collaboration with the local community, John Maine sculpted a series of five terraces at Chiswell, Dorset. These draw together both historical and geological aspects of the local landscape. The terraces reflect the strip lynchet field systems of Dorset and the stratigraphy of the Portland Beds, the stone of which is used within the sculpture. Peter Randall-Page created a series of giant snails from Purbeck marble, highlighting the fossils commonly found within the rock of the Purbeck. These are located not within a visitor centre or on a commissioned sculpture trail, but lie, unmarked, along a cliff path, prompting the surprised onlookers to reflect upon the location, fabric and form of the object in front of them (Fig. 20.1).

The work by Andy Goldsworthy goes one stage further, as well as using natural materials, many of his forms are ephemeral and their destruction by weathering, water and time connects us to our surrounding landscape and natural processes. Activities modelled on this type of work, environmental art, are now common at many country parks and heritage centres throughout the country. Using the approach of both observation and participation, environmental art is ideal within an urban setting where the spatial extent of charismatic geological resources may be limited.

Fig. 20.1. Sculpture by Peter Randall-Page as part of the New Milestones Project, Dorset. (Photography courtesy of: Common Ground.)

The urban wildlife partnership

Local geological societies, the BTCV and Common Ground together form three allies with whom we can increase awareness to further earth heritage conservation. It requires, however, effective communication. The Urban Wildlife Partnership (UWP) is an established medium through which links may be forged to promote earth heritage conservation to a wider audience. The UWP comprises about eighty bodies, including 51 urban wildlife groups, as well as local councils, Groundwork Trusts and wildlife trusts. It involves 17 500 people, with around 2000 of these as active volunteers.

The key objectives of the UWP are: to facilitate the exchange of ideas and experiences about urban wildlife; to establish support from grant aiding and sponsoring agencies for all types of urban wildlife intiatives; to encourage action; to raise awareness within other organizations, groups and individuals, including the business community, about the need for nature conservation in urban areas; to promote an understanding of urban wildlife and its links with sustainability through formal and informal education; and to encourage the development of policies within local and central government to care for and enhance urban wildlife.

The UWP forms a network which has the potential to involve a large number of people in the urban environment with earth heritage conservation ranging from local people wishing to enhance the nature conservation value of their green space to those who formulate policy within local councils. Using the UWP, earth heritage conservation can reach a wide cross-section of the urban community and provide a solid route to integrate earth heritage conservation into nature conservation plans and strategies .

Linking in to such existing urban initiatives may be of direct benefit. For example, only a few hundred metres from Trehafod, mid-Glamorgan, is a country park containing two waterfalls resulting from the original cutting of the Barry Sidings, now developed as a cycle/footpath running between Trehafod and Pontypridd. The waterfalls expose excellent sections of Pennant Sandstone and Upper Coal Measures, as well as overlying silts, peat and till. An interpretation centre is planned for the area, and although those concerned with its development are aware of the geology, it is not at the top of their agenda. Attitudes such as this may not be changed overnight, but the importance of the local geology may have been considered more fully had the local authority been aware of earth heritage conservation via the UWP network.

Conclusion

The urban environment has been developed for people by people. Its form reflects its cultural and social history, with a more remote connection to the underlying geology and local landforms. The importance of the inhabitants of towns and cities in any nature conservation strategy therefore cannot be understated. Numerous links can be forged with other pre-existing groups to further the role of earth heritage conservation in urban areas. These groups focus largely on *people* and their interaction with their natural environment. Those with an interest in geological conservation tend to get absorbed with the physical resource. For earth heritage conservation to be effective in urban areas, we must focus on the *people* as well as the features on their doorsteps.

References

Cutler, A. 1994. Local conservation and the role of the regional geological society. *In*: O'Halloran, D., Green, C., Harley, M., Stanley, M. & Knill, J. (eds) *Geological and Landscape Conservation.* Geological Society, London, 353-363.

---- 1996. The role of the regional geological society in urban geological conservation. *This volume.*

Keene, P. 1994. Conservation through on-site interpretation for a public audience. *In*: O'Halloran, D., Green, C., Harley, M., Stanley, M. & Knill, J. (eds) *Geological and Landscape Conservation.* Geological Society, London, 407-41.

McAdam, A. D. 1994. *Edinburgh: a landscape fashioned by geology.* Scottish National Heritage, Perth.

21 The role of the regional geological society in urban geological conservation

Alan Cutler

Summary

- The local geological society has a central role to play in raising public awareness of urban geology.
- This contribution reviews the activities and impact that the local geological society can have, from the perspective of a successful local group, the Black Country Geological Society.

This chapter explores the role of the regional geological society in urban geological conservation through the example of the Black Country Geological Society (BCGS). The BCGS has been actively involved in conservation since its inception in 1975, and has achieved significant success in the process of site protection in the Black Country.

Background to the BCGS

The Society's principal aims are:

1. To provide a programme of activities, lectures, field meetings, excursions and social events to cater for the continuing interest of members. Whilst some members are only interested in entertainment, their support and the activities described give the society a stable foundation, an essential framework and a public face without which the conservation effort would not be sustainable.
2. To try to counter threats posed to the well-being of local geological sites and thereby protect them.

Activities are reported in local newspapers and members regularly give interviews on both local and national radio. The BCGS guides walks on summer Sundays, and on a more formal basis, conducts field meetings for visiting parties and local 'A' level teachers. The Society mounts displays at selected local exhibitions and conferences and has played a major role in the organization of the well known and highly succesful Dudley Rock and Fossil Fair. We support major scientific conferences such as the Geological Curators Group (1985) and the Murchison International Silurian Symposium (1989). All of these activities promote the Society, but have their greatest impact in the promotion of the geology of the region. The Black Country and its geological heritage is now considered significant for tourism, and this has raised the profile of geological conservation both regionally and nationally (Cutler 1994).

Local site conservation in the Black Country

The BCGS has had a significant impact in protecting the urban geological resource in the Black Country. The history of local site conservation in the Black Country has been documented elsewhere (Box & Cutler 1988) and although the society was consulted on many planning applications from 1977 onwards, formal recognition of geological sites was not achieved until 1988.

Non-statutory wildlife sites in West Midlands County were identified by the Nature Conservancy Council, now English Nature, in the late 1970s and received the designation Sites of Importance for Nature Conservation (SINC). These sites were selected on biological criteria alone, and were subsequently formalized by inclusion in the County Structure Plan. Geology was considered only in a footnote to the schedule of sites, which referred to a separate list of geological sites held by the BCGS. This indicated at least a token interest in non-statutory earth science sites but in practical terms geological sites were never formally identified or recognized, and were, in effect, marginalized. Following the demise of the West Midlands County Council in 1984, the seven metropolitan boroughs became unitary authorities but SINC recognition was maintained, fortunately, through various local plans such as the Recreation and Open Space Subject Plan in Dudley. The first geological SINCs in the Black Country boroughs were added to the schedules in 1988 following recommendation and justification by the BCGS.

The Town and Country Planning Act (1990) required each of the metropolitan boroughs to produce its own Unitary Development Plan (UDP), the strategic policy framework by which development and land-use can proceed and indeed controlled (Walsall MBC 1991; Sandwell MBC 1992; Dudley MBC 1993; Wolverhampton MBC 1993). The policy towards SINCs as set out in each UDP is significant and states, *inter alia*, that 'Development will not normally be allowed where it would unreasonably prejudice the nature conservation value of designated Sites of Importance for Nature Conservation'. Whilst this may not afford statutory protection as is the case with Sites of Special Scientific Interest (SSSI), the UDPs are the sole statutory land-use plans for the boroughs and therefore confer a high level of protection on designated sites.

The BCGS has contributed significantly towards the acceptance of geological conservation in the region and continues to do so. The prioirtiy for the BCGS has primarily been site identification and recording, and promotion of the need for gelogical site conservation. Ultimately this has led to the acceptance of the need to protect our earth heritage and to formal recognition of selected geological sites by the four metropolitan boroughs that constitute the Black Country, and by Birmingham City Council. Site maintenance or interpretation has received comparatively little attention but will become increasingly important in the long term.

The Black Country Nature Conservation Strategy

The UDP policies can only address land-use issues and therefore a Nature Conservation Strategy is needed to address the wider issues involved. In 1988 the four Black Country boroughs, Dudley, Sandwell, Walsall and Wolverhampton, and the Black Country Development Corporation commissioned the Urban Wildlife Trust to write a Nature Conservation Strategy for the Black Country. The strategy, published in 1994 (Black Country MBCs & English Nature 1994), is a policy document which sets out the overall approach of the four boroughs and the Development Corporation to the conservation of the Black Country's natural resource. The Nature Conservation Strategy has no legal standing but provides planning guidance on a range of nature conservation issues related to the policies of the boroughs' Unitary Development Plans. Four major themes have been identified as the cornerstones of the strategy's overall framework: (1) protection of the nature conservation resource; (2) enhancement of the nature conservation resource; (3) people and nature conservation; and (4) nature conservation - the wider implications. Another feature of the BCNCS is the formal recognition of 'third tier sites' which are designated as Sites of Local Importance for Nature Conservation (SLINC). SLINC sites by definition are not of such high quality as SINCs but broadly speaking will possess social rather than scientific value. Given the authorship of the strategy there is an understandable bias towards flora and fauna. The BCGS, with support from other wildlife organizations, had an input during the initial draft phase and it is clear that the authors in conjunction with the Local Authorities Steering Committee subsequently endeavoured to incorporate geology in a sincere and comprehensive way. This is illustrated by the following extracts of principles taken from the strategy.

- The boroughs and the Development Corporation will strictly apply the site protection principles set out in this strategy to themselves. They will set an example to the private sector and other organizations by protecting wildlife and geological features on their own land from damage or destruction.
- The boroughs and the Development Corporation will set an example to others by expertly managing wildlife and geological sites owned by them and by encouraging appropriate management on privately owned sites.
- The boroughs and the Development Corporation will seek to enhance the Black Country's natural resource by habitat creation and revealing geological features.
- The boroughs will seek to ensure that wildlife and geological sites are regularly monitored and that the data on the information base are kept up-to-date.
- The boroughs will continue to publicize and promote wildlife and geological issues and cooperate with others in the dissemination of information on nature conservation in the Black Country.

The BCNCS is a significant advance for geological conservation and the BCGS has had a major impact in the inclusion of the earth sciences within it. Future commentators will assess its influence. Already Dudley Metropolitan Borough recognizes 'it is important to realise that the production of the BCNCS is not seen as an end in itself, but as a step in the process of making Dudley and the Black Country a better place for Nature Conservation', and is now in the process of formulating an 'Action Plan' in which geology is top of the list of main topics.

Geology and wildlife

BCGS experience shows that developing links with its wildlife counterparts is not only productive but essential for geological conservation in the long term. This does not imply that geological conservation will be subsumed within the wildlife movement. Indeed the special needs of geology are clearly recognized as indicated by initiatives such as a Geological Code of Practice currently being developed in the region (Reid 1996). There are also signs that the County Trusts are rapidly moving away from their original general nature conservation position to one that is more specifically wildlife orientated. Nevertheless, at the strategic level it is logical that nature conservation and the natural environment should be treated in their widest sense. The establishment of Local Nature Reserves (LNRs) with any geological interest will normally require substantial cooperation between geological and wildlife conservationists. LNRs are highly attractive in the urban context and, significantly, offer statutory protection.

Conclusion

There are many ways in which a regional geological society can help to raise the profile and pursue the cause of geological conservation, particularly in urban areas. Several have been highlighted in this paper through the long involvement of the BCGS in urban geological conservation. The society's involvement with the geological collection at Dudley Museum (Cutler 1994) has also yielded dividends and has been successful in developing and promoting awareness of geological heritage. In urban areas other facets such as churchyard gravestones, local stone walls, kerb stones and buildings are also significant in the overall theme of geological conservation and have an important bearing on the character of a region, its heritage and in understanding the role that geology plays in the human experience. The long term success of non-statutory site protection needs public support, which will be much more forthcoming where geology is recognized and appreciated and seen to be an integral part of the natural environment.

References

Black Country MBCs & English Nature. 1994. *Black Country Nature Conservation Strategy.* English Nature, Sandwell MBC, Dudley MBC, Walsall MBC and Wolverhampton MBC.

Box, J.& Cutler, A. 1988. Geological Conservation in the West Midlands. *Earth Science Conservation*, **25**, 29-35.

Cutler, A. 1994. Local Conservation and the role of the regional geological society, *In:* O'Halloran, D., Green, C., Harley, M., Stanley, M. & Knill, J. (eds), *Geological and Landscape Conservation*. Geological Society, London, 353-357

Dudley MBC. 1993. *Unitary Development Plan.*

Reid, C. G. R. 1996. A code of practice for geology and development in the urban environmnet: a new local authority initiative. *This volume.*

Sandwell MBC. 1992. *Unitary Development Plan* (Draft for Deposit)

Walsall MBC. 1991. *Unitary Development Plan* (Draft for Deposit)

Wolverhampton MBC. 1993. *Unitary Development Plan.*

22 Geotourism, or can tourists become casual rock hounds?

Thomas A. Hose

Summary

- Site interpretation plays an important part in raising interest in the conservation of geological sites.
- Geotourism is the provision of interpretative and service facilities which enable visitors to gain some knowledge and understanding of geology.
- Increasing the impact of geotourism will increase the awareness of the need for the conservation of geological sites.
- The results of several surveys of the vistors to geologial sites are presented and the implications for site interpretation considered.

This chapter summarizes a theoretical framework for research on geology and tourism - geotourism - and the results of early visitor studies undertaken at interpreted earth heritage sites within England. In presenting it the author is keenly aware of the disparate perceptions and needs of at least two distinct groups of likely natural heritage researchers and managers. These are: professional earth scientists generally interested in the academic, educational and research value of a site, but with little or no knowledge and/or interest in either more popular approaches or tourism; and heritage and conservation managers with a limited knowledge of earth science conservation and with some understanding of the tourism industry and of attractions management.

Geology and tourism: some basic issues

Earth sciences and tourism studies: a worthy conundrum?

In view of the above, much of the material referenced in this study has been restricted to that which is considered readily available and accessible to a wide readership. For those earth scientists sceptical of any linkage between earth heritage and tourism it is worth noting that: '. . . 30% of foreign visitors cite our architectural heritage as the sole reason for visiting the United Kingdom, and 70% include it as one of their reasons; in fact, 75% make it to a historic site during their stay. Heritage is a multi-million pound industry and its contribution to the success of tourism in the United Kingdom is undoubted; like it or not, many visitors come to this country because of . . . the thatched cottages, the mediaeval churches and (increasingly) the satanic mills. The purists may not like it. The cynics may sneer at it. But at the end of the day there is a powerful argument for protecting the heritage because it earns its own living' (Ross 1991, p. 175).

Promoting various heritages

The developing interest in the United Kingdom's archaeological, architectural and industrial heritage and associated packaged tourism, has often resulted in the loss of mine tips, spoil heaps, quarries and pits to amenity landscaping as the old industrial landscapes and townscapes are remodelled into tourist-acceptable photogenic backdrops. Cultural heritage conservation has frequently led to the modification/refusal of planning permission for numerous extractive industry and road construction proposals, but all too frequently the geotourism potential of these is obscured (Baird 1994). The concept of heritage management is now well accepted and: 'Much thinking about heritage, from those who talk about preservation for posterity to those who talk about producing and marketing a product, starts with finding or making a resource and prepares it for predetermined uses. But it is appropriate to view the process of heritage management as arising from the formation of these uses in relation to anticipated or expressed desires and needs, and to look at the work which is done in constituting the heritage resource as part of a service to the public or a more specifically defined constituency. The quality of constituency-building is probably more important than the resource in determining the character of the result.' (Alfrey & Putnam 1991, p. 87).

Earth heritage and the professional earth science community

Noting the success of the other heritage and tourism lobbies, it is a pity that far too many academic and professional earth scientists, despite at least thirty years of debate, still 'stand idly by, hiding our major discoveries . . . and the excitement that goes with them' (Baird 1968, p. 223) whilst other scientists achieve recognition for their achievements. It might well be that as long recognized by archaeologists and museum staff the idea that 'good on-site presentation . . . is good for raising awareness, public relations, and for generating income and support for continued work' (Binks *et al.* 1988, p. 2) is slowly gaining ground at least with the earth heritage conservation community if not those engaged within the extractive industry (O'Halloran *et al.* 1994). Within the former the role of museum staff (Stanley 1992; Knell 1996) has been crucial. Given that the role of interpretation is 'to assist in conservation, which itself is aimed at ensuring that future generations will be able to enjoy the nation's natural and historic heritage' (Aldridge 1975, p. 6), an earth heritage interpretive strategy could well generate the public pressure required for its promotion and protection. Clearly, the first task in establishing geotourism is to define its framework and outline its physical basis.

Urban earth heritage sites as geotourist attractions

With some exceptions (Rees & Harris 1973) the traditional emphasis for earth science fieldwork has been a rural one. However, 'Britain is perhaps the most urbanised nation in the world with approximately 80% of her population living in urban areas. While the process of urbanisation, defined as any increase in the proportion of total population living in urban administrative areas, has slowed down, projected population growth predicts a continued expansion of towns and cities in England and Wales. Although higher real incomes and greater mobility will probably augment the tendency to seek recreation beyond the city, technology is unlikely to produce, in the foreseeable future, a system that will enable the majority of city-dwellers to leave the city for daily and regular weekend recreation. Opportunities must continue to be provided for outdoor recreation within the urban milieu . .' (Balmer 1971, p. 113).

As a result of this urbanization, particularly from central England northwards, many towns and cities now encompass a range of permanent and ephemeral earth heritage sites with some potential for tourism development.

The nature of the urban geotourism resource varies from museums (e.g. Dudley Museum and Ludlow Museum) and interpretative centres (e.g. Fossil Grove, Glasgow) to civil engineering projects (especially canal, road and railway cuttings and tunnels such as Dudley Tunnel or Black Country Museum), buildings (e.g. *A Geological Walk Around Dorchester*, Dorset Natural History and Archaeology Society) and disused extractive industry sites (e.g. Salthill Quarry Geology Trail, Clitheroe, Lancashire) and natural landform features (e.g. Castle Rock, Edinburgh, which has interpretative panels provided by the Edinburgh Geological Society in West Prince's Street Gardens). Urban sites offer unique opportunities to examine rocks, including the effects of weathering (Wright 1986) juxtaposed from widely dispersed locations and facies; numerous urban geology trails, often using cemeteries, exploit this resource (e.g. Perkins 1984; Robinson 1984, 1985).

The recognition (Reid 1994) and promotion of these and new sites is all the more urgent given the sanitization of the industrial and urban landscapes of the past two decades. In part, their situation is not unlike that of the Victorian industrial and architectural legacy savaged by redevelopment in the 1950s and 1960s, until its heritage potential was recognized and promoted. Success here has been because: 'Those who have succeeded in extending the boundary of heritage concern to include the remains of industrial civilization have had to demonstrate the cultural benefits of this 'new' heritage and have often been innovative in developing new programmes or cultivating new constituencies. They have accomplished some quite extraordinary things; a generation ago industrial heritage was virtually unknown; now it has become something to conjure with-even an element in the regeneration of areas devastated by the decline of key industries.' (Alfrey & Putnam 1991, p. 2).

It is that level of demonstration of cultural benefits that earth scientists must apply to conserve and promote the nation's earth heritage. In this, an understanding of the interrelationships of heritage and tourism, especially the nature of the tourist, would seem essential. Unfortunately, there is little published work on visitors to United Kingdom earth heritage sites. This is not a singularly British shortcoming for little has been published elsewhere (Patterson & Bitgood 1986) on purely earth heritage (as opposed to museum and dinosaur-centred) attractions: 'Despite the importance of fossicking as a tourist attraction in New England region, relatively little research has been directed towards the various aspects of its development and promotion. In particular, there had been no previous research on the motivations and management requirements of fossickers.' (Jenkins 1992, p. 131).

Recent work in the United Kingdom (Burek & Davies 1994; Hose 1994 *a*, *b*) has sought to provide such information. It now seems opportune to develop a framework for the development of research on tourism centred on our earth heritage.

Geotourism and its clients

Geotourism defined

Geotourism offers both a new packaged tourism product and the potential to constituency-build for the earth sciences and earth heritage conservation (Besterman 1988; Keene 1994). It can be defined as: 'The provision of interpretative and service facilities to enable tourists to acquire knowledge and understanding of the geology and geomorphology of a site (including its contribution to the development of the earth sciences) beyond the level of mere aesthetic appreciation'. Therefore, it encompasses the life, work, publications, workplace, residences and final resting places of earth scientists, together with an examination and understanding of the physical basis of earth heritage sites. For the purposes of this framework tourists are defined as persons experiencing a leisure activity away from their normal place of permanent residence. Geotourism, unlike many other forms of tourism, is not limited by the seasons; it has the potential to extend the nature and timing of tourism provision. It also offers an alternative attraction to relieve visitor pressure on other natural and cultural heritage tourist sites.

Geotourism and tourist behaviour

Approaches to understanding tourist behaviour have frequently resulted in the development of models of tourist satisfaction and tourist typologies; a few with some relevance to geotourism and urban tourism have been published in Canada and the USA. In general terms some five considerations (Ryan 1995) have been thought important by those developing tourist satisfaction models. Those significant for geotourism are: (1) the perceived importance of the activity: ego, self-development, self-enhancement, meeting perceived roles, and responding to perceived requirements of other significant social groups and individuals; and (2) the importance of the activity being evaluated not only by need, but also by expected outcomes: questions pertaining to perceptions of both needs and outcome have to be answered.

One component of a four-point leisure motivation scale (Beard & Ragheb 1983) is worth noting; the intellectual component, which assesses an individual's motivation to participate in leisure activities that involve learning, exploring, discovering, thought or imagination. Generally, research on tourist motivation and satisfaction has two strands. The concept of a leisure/tourism career (Pearce 1988) suggests that individuals' choices are governed by the assimilation of prior experiences resulting in a near linear progression of new experiences and activities. Alternatively, tourist satisfaction and holiday choice might be strongly dependent on repeating prior experiences, the ingraining of prior experience (Laing 1987) which gave satisfaction and/or security. In essence, tourists are creatures of habit and do not necessarily respond to changes in technology or available experience. The latter probably has serious implications for new packages such as geotourism.

Geotourism and tourist typologies

In the USA 'Values, Attitudes and Lifestyles' (VALS) centred research has produced an eight-category tourist typology which includes strugglers, believers, strivers, makers, fulfillers, achievers, experiencers and actualizers. A similar United Kingdom six-category typology, 'Outlook', lists: trendies, pleasure seekers, indifferents, working class puritans, sociable spenders and moralists. Similar typologies for earth heritage visitors have been proposed in the United Kingdom. Three categories of palaeontological sites' potential users, equally applicable to general earth heritage sites, have been identified (Besterman 1988; Clemens 1988).

1. Recreational: amateurs (children and adults) as individuals or parties, whose idea of a good day out is to go and look at or for fossils, with a broad expertise range; from absolute beginner (for whom the approach

in Rees & Harris (1973) is still valuable) to lifelong enthusiast (for whom the information in Lichter (1993) is valuable) with a specialist knowledge equalling, or even exceeding, that of academic specialists.

2. Educational: a broad range of pre-school to post-graduate users with quite different needs but similar access requirements. For them, earth science is essentially a field subject and serious students must examine many sites as part of their training.
3. Commercial: collectors, often with a specialist knowledge, equalling or exceeding that of academic specialists, working either full or part-time. Unfortunately, some have a preference for 'perfect' specimens and may discard much material that would be found useful by individuals in the other categories in their quest for readily saleable items.

Geotourism's client base will be drawn from mainly the recreational users. Of course, the educational users could also take advantage of any such provision. Keene (1994) cautions that: 'Whatever one's views about earth science education for the masses, there is clearly a yawning gulf between the educator and the public - a polarity which seems painfully more apparent in the earth sciences than in other environmental disciplines such as archaeology, ecology or history.' (Keene 1994, p. 408). He identifies four potential target audiences, each with their own needs, which, with some modification, can be summarized as follows.

1. Education groups; from schools, colleges, universities and adult education organizations; as such, individuals are accustomed to a structured, linear learning environment.
2. Interested information seeking adult non-specialists; an articulate, and growing, group of individuals predisposed to be sympathetic to conservation and geological sites; its members are avid readers of associated literature.
3. Thoughtful adult non-information seekers; an articulate group more interested in the experience, rather than the meaning, of geological sites.
4. Mass of general public; generally disinterested and non-readers for whom visiting geological sites is likely to be mainly an accidental and probably social event. The vast bulk of the population.

Geotourism: the packaged product

Interpretation and conservation

Whichever way the users of geological sites are categorized, their basic needs are similar. Published guides, pitched at an appropriate level, are essential; for example, a guide to the fossils of a site could refer to a specific marine, squid-like animal, with a coiled shell as 'an ammonite' for the geotourist and as *Deshayesites vectensis* for the advanced enthusiast/educational geologist. Perhaps, herein lies geotourism's greatest challenge. Earth heritage nomenclature, even for specialists, is particularly difficult to assimilate. There is some merit in using simple common names such as in the case of garden and wild flowers (Dony 1986) and birds (Gruson 1976); however, it has to be accepted that many keen gardeners do become adept at recalling the taxonomic names of familiar plants! Undoubtedly, increased familiarity with earth heritage material will help meet this challenge. Fortunately, little specialist equipment is required by the geotourist, a hand lens/magnifying glass often being of more value than a geological hammer. Indeed, geotourists should be actively discouraged, in the interests of conservation and for personal safety from carrying and using hammers. Such an approach is very much in line with conservation developments in amateur field botany (where the camera and recording form has replaced the vasculum and subsequent herbarium sheet specimens) and ornithology (where the sketch pad and camera have replaced the egg and skin collections) as the prime means of enjoying and recording past experiences.

Provision categorized

Geotourism interpretative provision can take a number of forms: personal service - lectures and guides; publications - leaflets, postcards, posters and booklets; trails - self-guided either by waymarked leaflet booklets and panels or audio tapes; or display media - on-site panels, interpretative exhibitions, museums and visitor centres.

Of course, the various media can be combined. Their theoretical basis has been widely examined (e.g. Pennyfarther 1975; Piersenne 1985) since the seminal American outdoor recreation model (Tilden 1967) was published. Patmore (1970, p. 75) points out that: 'By any standards, England and Wales, are the most urbanised of countries. It is hard to distinguish between town and country dwellers in any meaningful way when the two are so closely intermingled and related. . . although we are a nation of townspeople (indeed of city-dwellers, for one person in three lives within the immediate confines of the six great connurbations alone), towns still cover a comparatively small proportion of the land as a whole.' The traditional countryside approaches to earth heritage interpretation are clearly still valid within the urban context. Trail guides (e.g. Jenkinson 1991) and leaflets are the least visually intrusive and most cost-effective interpretative media. Conceivably, tourism operators could be encouraged to distribute the leaflets and visit some of the sites as part of their excursion programmes. For these, the printing of appropriately captioned postcards (briefly and simply describing the geology of the view) and the manufacture of souvenir items (especially jewellery and giftware) based on local geological material could address a potentially profitable market. In the interests of geological conservation, replica fossils could be manufactured for retail; these also have the advantage that perfect specimens, rather than the usual, can be used to produce the moulds from which replicas can be cast.

Resourcing geotourism

Given its diverse nature, geotourism could be promoted and supported in a number of ways. It could be resourced through conservation and sustainable tourism measures. Again, it could be funded by local and regional tourism strategies. Whilst in the past various publicly-funded agencies have been at the fore in the provision of materials that can nestle under the newly defined banner of geotourism, increasingly there is scope for commercial sector provision (De Bastion 1994). The commercial producer must of necessity focus on the popular and create populist material, likely to retail successfully. The publicly-funded producer will most likely focus on educational potential. Geotourism, however, will always use a variety of interpretative media.

Geology and tourism: visitor studies

Visitors to three contrasting earth heritage attractions have been surveyed to obtain basic psychographical data, together with an assessment of the effectiveness of interpretative provision. Unfortunately, it is still too early in the research programme to examine accurately the relevance of the various tourist typologies currently promoted. Low-level data analysis, however, has been applied to generate information on the current state of British geotourism which is presented in a series of appended summary figures and tables.

The National Stone Centre

Location and issues

The National Stone Centre, which opened in 1989, is near Wirksworth, Derbyshire some 200 km north of London and just south of the Peak District National Park. It aims to 'tell the 'Story of Stone' in the United Kingdom, its formation, industrial history, uses and related environmental issues. These topics are the subject of the indoor exhibition (Thomas & Hughes 1993, p. 17) which is a purely earth heritage attraction.

The two surveys

The Discovery Building, housing the 'Story of Stone' exhibition, shop and small cafe, is the site's only permanent shelter and was the sampling station during 7-16 August 1993. Two surveys were undertaken. A respondent-completed questionnaire was issued to all adult parties purchasing tickets for the 'Story of Stone' exhibition; some fifty (48% response rate) were returned. Over eight days (chosen to avoid special events) subjects leaving the building (between 11.00 and 13.00 and 14.00 and 16.00) were interviewed; 76 interviews (92% response rate) were concluded. Some 124 questionnaires were available for analysis.

Visitor characteristics unearthed

A high proportion of respondents were first-time visitors (Fig. 22.1). The most populous group were those in mid life-cycle with families. Those aged 30-64 years were generally visiting either as couples or in small family groups. The 19-29 years age group were under-represented in the survey suggesting it is perceived as of little interest to them (Fig. 22.2). Couples, at just over one third, are a major component. Groups of

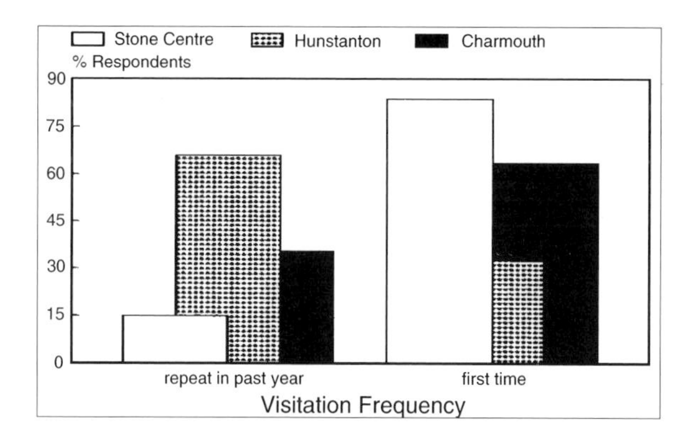

Fig. 22.1. Site visitation of all respondents.

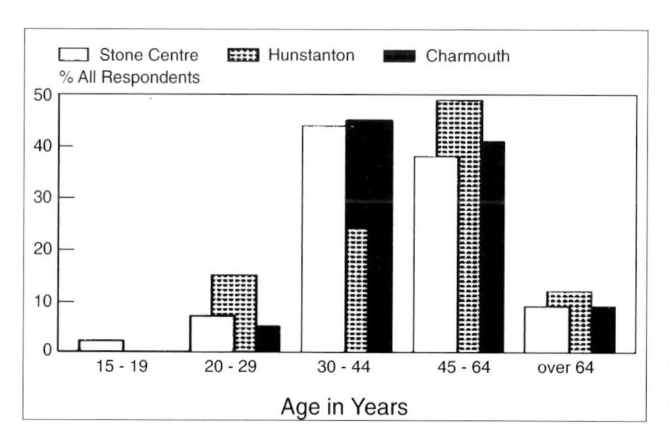

Fig. 22.2. Age profile of all respondents.

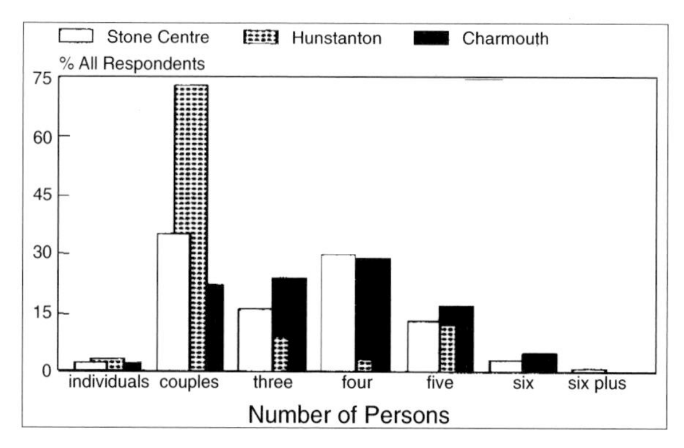

Fig. 22.3. Party size of all respondents.

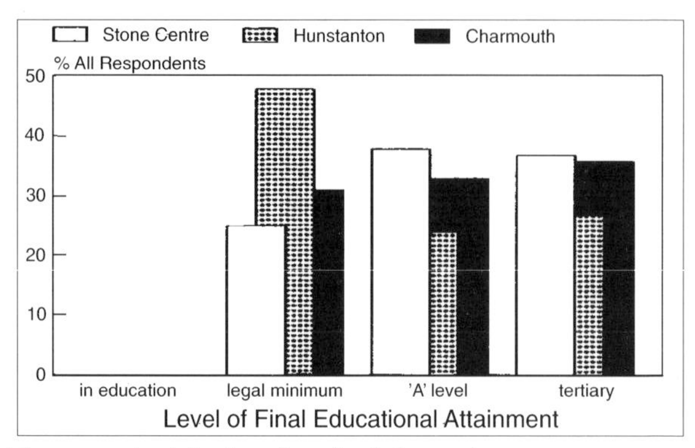

Fig. 22.4. All respondents' educational attainment.

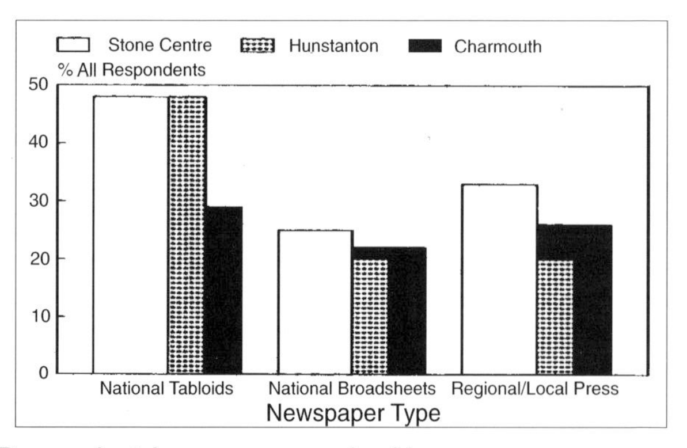

Fig. 22.5. Respondents' newspaper readership.

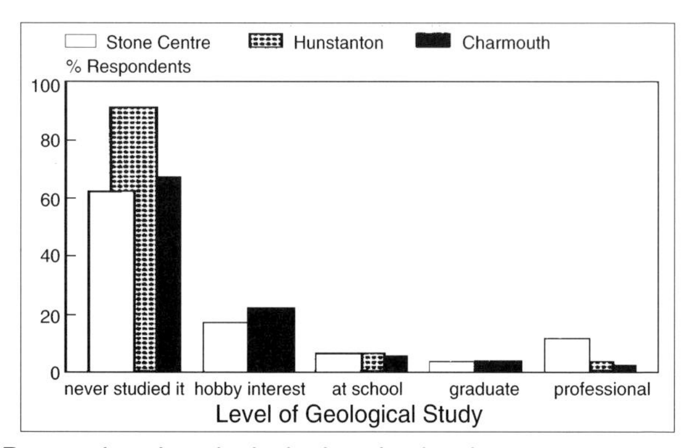

Fig. 22.6. Respondents' geological education level.

three persons accounted for about a fifth of group visits. Small family groups accounted for about half of all visits (Fig. 22.3). The high percentage educated at tertiary level suggests a relatively large professional component in the visitor population (Fig. 22.4), although this is at variance with the press data (Fig. 22.5). The high readership levels of tabloid and regional newspapers is indicative of the social (low income, low educational attainment) and geographical (that is, northern England) origins of many respondents. A significant minority of respondents, probably linked to the regional significance of coal-mining and quarrying, claim some earth science interest and knowledge (Fig. 22.6).

Interpretative effectiveness

As part of the survey two illustrated prompt cards were used. One was on common British fossils: the illustrations were chosen to represent those likely to be found on-site, in the Centre's literature and elsewhere; although respondents often used their common names, recall rates were quite high for some fairly distinctive fossils. The other was a modified (to clarify the points in question) representation of the site features diagram, which superimposed the modern quarry view over a reconstruction of the site some 320 million years ago, found at various points on the geology trail; responses to it indicated respondents' understanding of the site. Unfortunately, few respondents (17% of those interviewed) had actually walked the trail, which tended to be visitors' last on-site activity, prior to their interview. However, of those who had walked the geology trail, a number of respondents thought the quarry itself was the actual lagoon shown in the reconstruction, rather than the site of its deposition products. Recall rates were generally fair (between 15% and 46%) for such technical knowledge (Figs 22.7 & 22.8). The ability to name one particular geological system is related to a successful film, *Jurassic Park*, rather than earth science understanding (Fig. 22.7).

Other considerations

Informal observation indicated that visitors preferred to purchase non-printed matter. Many visitors main purpose in entering the site was to purchase jewellery and giftware from the shop. The exhibition was fairly popular and the geology trail was the most used on-site facility.

Hunstanton Cliffs SSSI

Location and issues

Hunstanton Cliffs, with their red and white Chalk, form a prominent low coastal feature adjacent to a beach, promenade and funfair, in a small Norfolk seaside resort some 160 km northeast of London. A single interpretative Earth heritage panel was erected by English Nature in 1993. The potential annual audience was estimated (Page 1992) at some 1000-10 000 consisting of casual passers-by and interested visitors. Hunstanton is considered to be a basically recreational, rather than earth heritage site.

The two surveys

In May 1994, over 10 hours, a visitor survey was undertaken (at various stages of the tide and with the weather being unusually cold and wet for the season) on three successive Sundays (to ensure that genuine seaside visitors rather than shoppers were encountered). A log was kept of visitors' response to the panel. Interviews were conducted with those viewing the panel in excess of 30 seconds (although good use of the panel was judged to require at least one minute). As part of these, interviewees were shown two illustrated prompt cards, one of which concerned common British fossils. Thirty-two interviews (an 80% response rate) were concluded.

General visitor characteristics

The high percentage of couples (Fig. 22.3), those 45 years and older (Fig. 22.3) and the relative paucity of first-time visitors (Fig. 22.1) reflects the popularity of the area with retirees and also coach outings. Again, it is much used as a beach resort by families from the nearby inland towns. The generally low level of educational attainment (Fig. 22.4) reflects both the age profile and the local (still mainly agricultural) employment pattern of the area. The high readership levels of tabloids and low regional newspaper readership (Fig 22.5) is indicative of the respondents' social (low income, low educational attainment) and geographical origins (East Anglia and the Midlands).

The cliffs discovered?

The high non-response rate (76%) to the panel (Fig 22.9) is indicative of respondents' general educational attainment and desire to partake of the usual seaside activities. Response rates rose as high tide approached, forcing people off the sandy beach and onto either the promenade or the thin strip of beach beneath the cliffs. The unseasonable inclement weather was also undoubtedly a viewing time factor. The maximum recorded viewing time was 2.51 minutes; the minimum was 0.05 minutes and the mean was 1.02 minutes. Media hype has clearly influenced the ability to name geological systems (Fig 22.7). The generally low accuracy in response rate for fossils (Fig. 22.8), except for the easily recognizable bivalve mollusc (very similar to those found living on the foreshore), suggests that the public can recall such material from drawings; their knowledge shortfall probably lies in a lack of exposure to fossils in an area lacking hard rock sites (i.e. Norfolk). Some degree of the panel's interpretive success can be noted; 42% of respondents recalled the area was 'like the Bahamas' and 27% 'a warm, clear tropical sea' when the rocks forming the cliff were made. Some 53% of respondents gave an accurate indication of the age of the rocks forming the cliffs. Clearly, the use of common terms and colourful word pictures for earth heritage interpretation were helpful for general recreational tourists at this site.

Charmouth Heritage Coast Centre

Location and issues

Charmouth, a quiet seaside village in Dorset, lies some 200 km southwest of London. Adjacent to a car park, an earth heritage interpretative panel (associated with two on local wildlife and a location map) and visitor centre have been provided; the latter opened in 1985 (with displays installed in 1986) and has free admission with some 30-35 000 visitors annually. In 1993 an audio-visual facility (for a modest charge) was opened. The Centre is on the sea front on the first floor of a stone-built old cement factory; underneath are a commercially operated cafe, and geology and souvenir stores. Nearby are numerous coastal sections in tall, highly unstable cliffs of fossil-packed Jurassic (Liassic) dark shales, limestones and sandstones. It is both a recreational area and a popular earth heritage site.

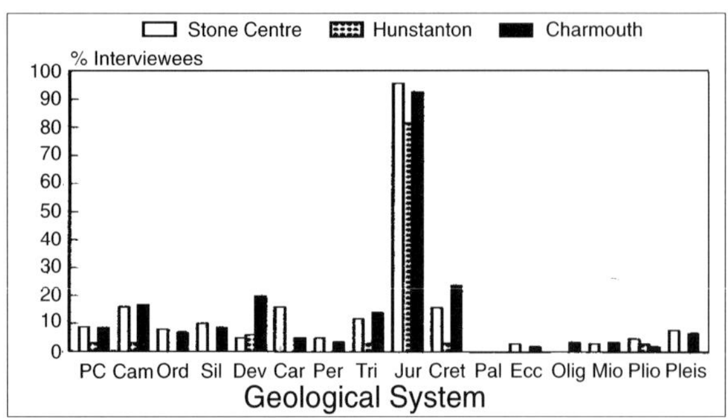

Fig. 22.7. Interviewees' geological knowledge gauged on their ability to name geological systems.

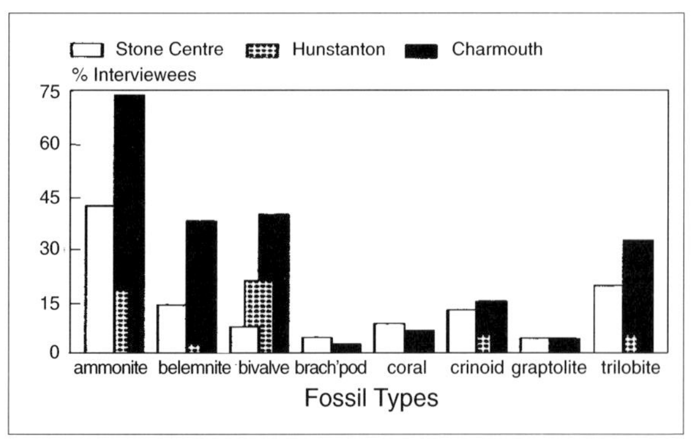

Fig. 22.8. Interviewees' common fossil recognition. Based on their response to a prompt card with simple drawings of common fossils on it.

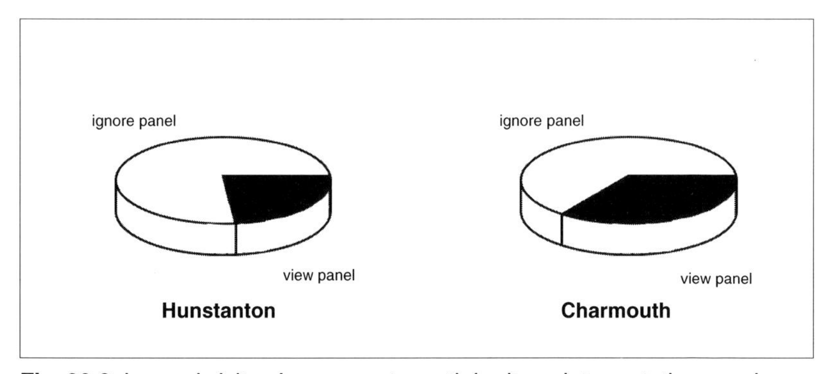

Fig. 22.9. Logged visitors' response to earth heritage interpretative panels.

The three surveys

Over three days in June 1994 over six hours were spent logging usage of the earth heritage interpretative panel. Interviews to assess its success were suspended when it became clear that it was viewed by very few respondents (Fig. 22.9)! From 21-27 August 1994 visitors using the Centre (between 11.00 and 13.00 and 14.00 and 16.00) were selected on a first past the post principle as they left the building and interviewed; some 58 (90% response rate) interviews were completed.

Facts unearthed

Interpretative panel usage was greatest in the afternoon, when it was cooler, and at high tide. The maximum viewing time logged was 4.36 minutes; the minimum was 0.03 minutes and the mean was 1.04 minutes. The high percentage of couples (Fig. 22.3) is indicative of those in the later stages of the life-cycle (Fig. 22.2) represented at the site and clearly seen to be visiting the Centre. The roughly equal split of repeat and first time visits (Fig. 22.1) probably reflects the large number of individually-owned local static caravan points and Charmouth's popularity with the population of the adjacent inland towns' population as a recreational beach. The high level of tertiary education (Fig 22.4) is noteworthy.

Other considerations

Informal observation evidenced the reluctance of the public to purchase even inexpensive geological publications; on two occasions relatively expensive (for unplanned) purchases of geological hammers (£12-50) were made, but leaflets costing 15p and 40p explaining where fossils could be found were declined. Unfortunately, the clear warnings, on both the earth heritage interpretative panel and the displays within the Heritage Coast Centre, on the dangers of collecting beneath the area's unstable cliffs and to collect (in the interests of conservation) only from loose blocks were ignored by numerous individuals.

Summing up the experience

Rockhounds considered

The three sites display some evidence of regional variation in client base, however, broad trends, especially from informal observation studies, are discernible. The vast bulk of visitors arrive in couples and small family groups. Most respondents were in the 30-64 years age range (Fig. 22.2) visiting with children of primary and lower secondary school age. Inexpensive souvenirs and moderately-priced giftware and, given the right setting, prepared fossils and mineral specimens can be profitably retailed to geotourists. Leaflets and specialized geological texts are marginal retail products. The readership levels of the tabloid press suggest clear limits to the vocabulary and style that can be used in interpretative material. Given exposure to earth heritage terms and materials respondents recall rates were encouraging (Fig 22.7 & 22.8). Respondents' general education levels seem to be higher at specifically earth heritage, rather than recreational sites (Fig. 22.4) and there is some correlation with their geological education (Fig. 22.6). For marketing purposes the readership levels of tabloids and the regional preference for non-national daily newspapers (Fig 22.5) is significant.

More rockhounds?

Despite the general lack of formal earth science tuition in schools there is some public interest in suitably presented earth heritage attractions. Some thought needs to be given to an appropriate vehicle for the earth heritage conservation message. Both the National Stone Centre and the Charmouth Heritage Coast Centre surveys indicate that a focused earth heritage site can be an attraction in its own right. The Hunstanton Cliffs and Charmouth panels surveys suggest that there is some public interest in earth heritage which, with appropriate interpretation provided separately from other information, can be memorable. Clearly there is potential for the development of such attractions, but this needs to be in the light of further and more detailed analyses of visitors to existing attractions. Importantly for geotourism: 'At the heart of tourism is the concept of travel; a chance to see new and strange sights, to learn about other places of the world, and to talk to others with different cultures and viewpoints.

. . For many people seeking a cultural return from their travel the educational opportunity is a real motivation for their touristic perambulation' (Ryan 1991, p. 27).

Leaving the last words to a geologist: 'We have a great opportunity to increase the prominence and respectability of our profession in the public eye, as materials take more skill to find and to refine, as engineering and technological feats dependent on geological advice become more common, an understanding of our whole environment, from food and health to oceans and earth origins, becomes vital. We stand now at a period of transition which offers us an opportunity to change our approach in presenting ourselves to the public.' (Baird 1968, p. 230).

Acknowledgements

The receipt of a research grant from the Buckinghamshire College and the cooperation and support of the staff at the Charmouth Heritage Coast Centre and the National Stone Centre is gratefully acknowledged.

References

Aldridge, D. 1975. *Guide to Countryside Interpretation: Part One: Principles of Interpretation and Interpretative Planning*. HMSO, London.

Alfrey, J. & Putnam, T. 1991. *The Industrial Heritage-managing resources and uses*. Routledge, London.

Baird, D. M. 1968. Geology in the Public Eye. *In:* Neale, E. R. W. (ed.) *The Earth sciences in Canada-a centennial appraisal and forecast*, Special Publications of Royal Society of Canada, **11**, University of Toronto Press.

Baird, W. J. 1994. Naked rock and the fear of exposure. *In*: O'Halloran, D., Green, C., Harley, M., Stanley, M. & Knill, J. (eds), *Geological and Landscape Conservation* . Geological Society, London, 335-336.

Balmer, K. 1971. Urban Space and Outdoor Recreation. *In*: Lavery, P. (ed.), *Recreational Geography*. David & Charles, Newton Abbot, 112-126.

Beard, J. G. & Ragheb, M. G. 1983. Measuring leisure motivation. *Journal of Leisure Research*, **15**, 219-228.

Besterman, T. P. 1988. The Meaning and Purpose of Palaeontological Site Conservation. *In*: Crowther, P. R. & Wimbledon, W. A. (eds)

The use and conservation of palaeontological sites. Palaeontological Association, Special Papers in Palaeontology, **40**, 9-19.
Binks, G., Dyke, J. & Dagnall, P. 1988. *Visitors Welcome*. English Heritage, HMSO, London.
Burek, C. V. & Davies, H. 1994. Communication of Earth science to the public - how successful has it been? *In*: O'Halloran, D., Green, C., Harley, M., Stanley, M. & Knill, J. (eds) *Geological and Landscape Conservation*. Geological Society, London, 483-486.
Clemens, W.A. 1988. Challenges of Management of Palaeontological Site Resources in the United States. *In*: Crowther, P. R. & Wimbledon, W. A. (eds) *The use and conservation of palaeontological sites*. Palaeontological Association, Special papers in Palaeontology, **40**, 173-180.
De Bastion, R. 1994. The private sector-threat or opportunity? *In*: O'Halloran, D., Green, C., Harley, M., Stanley, M. & Knill, J. (eds) *Geological and Landscape Conservation*. Geological Society, London, 391-395.
Dony, J. G. 1986. *English Names of Wild Flowers*. BSBI, London.
Gruson, E. S. 1976. *A Checklist of the Birds of the World*. Collins, London.
Herbert, D. T. 1989. Does Interpretation Help? *In*: Herbert, D. T., Prentice, R. C. & Thomas, C. J. (eds) *Heritage sites: strategies for marketing and development*, Avebury, Aldershot, 191-230.
Hose, T. A. 1994*a*. Telling the story of stone - assessing the client base. *In*: O'Halloran, D., Green, C., Harley, M., Stanley, M. & Knill, J. (eds) *Geological and Landscape Conservation*. Geological Society, London, 451-457.
---- 1994*b*. *Interpreting Geology at Hunstanton Cliffs SSSI Norfolk-a summative evaluation*. Unpublished paper, Buckinghamshire College, Buckinghamshire.
Jenkins, J. M. 1992. Fossickers and Rockhounds in Northern New South Wales. *In*: Weilerand, B. & Hall, C. M. (eds) *Special Interest Tourism*. Belhaven Press, London, 129-140.
Jenkinson, A. 1991. *Mortimer Forest Geology Trail*. Forestry Commission, Ludlow.
Keene, P. 1994. Conservation through on-site interpretation for a public audience. *In*: O'Halloran, D., Green, C., Harley, M., Stanley, M. & Knill, J. (eds) *Geological and Landscape Conservation*. Geological Society, London, 407-411.
Knell, S. 1996. Museums: a timeless urban resource for the geologist? *This volume*.

Laing, A. 1987. *The Package Holiday Participant, Choice and Behaviour.* PhD Thesis, Hull University.
Lichter, G. 1993. *Fossil Collector's Handbook.* Sterling, New York.
O'Halloran, D., Green, C., Harley, M., Stanley, M. & Knill, J. 1994. *Geological and Landscape Conservation.* Geological Society, London.
Page, K. N. 1992. *Information Boards for Geological and Geomorphological SSSIs.* Unpublished Research Report 24, English Nature, Peterborough.
Patmore, J. A. 1970. *Land and Leisure in England and Wales.* David & Charles, Newton Abbot.
Patterson, D. & Bitgood, S. 1986. *The Red Mountain Museum Road Cut: An Evaluation of Visitor Behaviour.* Center for Social Design, Jacksonville, Technical Report, **86-75**.
Pearce, P. L. 1988. *The Ulysses Factor: evaluating visitors in tourist settings.* Springer-Verlag, New York.
Pennyfarther, K. 1975. *Guide to countryside interpretation: part two: interpretative media and facilities.* HMSO, London.
Perkins, J. W. 1984. *The Building Stones of Cardiff.* University College Cardiff Press, Cardiff.
Piersenne, A. 1985. *Planning, scripting and siting panels, Environmental Interpretation, Centre for Environmental Interpretation.* Manchester Metropolitan University, Manchester.
Rees, G. & Harris, M. 1973. *Starting from Rocks.* Hart-Davis Educational, London.
Reid, C. 1994. Conservation, communication and the GIS: an urban case study. *In*: O'Halloran, D., Green, C., Harley, M., Stanley, M. & Knill, J. (eds)*Geological and Landscape Conservation.* Geological Society, London, 365-369.
Robinson, E. 1984. *London - illustrated geological walks. Book 1: the City.* Scottish Academic Press, Geologists' Association, Edinburgh.
---- 1985. *London - illustrated geological walks. Book 2: the City.* Scottish Academic Press, Geologists' Association, Edinburgh.
Ross, M. 1991. *Planning and the heritage - policy and procedures.* Spon, London.
Ryan, C. 1991. *Recreational tourism - a social science perspective.* Routledge, London.
---- 1995. *Researching tourist satisfaction - issues, concepts, problems.* Routledge, London.

Stanley, M. F. 1992. The National Scheme for Geological site Documentation. *In*: Erlkstad, L (ed.) *Earth science conservation in Europe - proceedings of the third meeting of the European working group on Earth science conservation.* Norsk Institute for Naturforskning, Troadheim.

Thomas, I. & Hughes, K. 1993. Reconstructing Ancient Environments. *Teaching Earth Sciences*, **18**, 17-19.

Tilden, F. 1967. *Interpreting our Heritage.* University of North Carolina Press, Carolina.

Wright, J. 1986. *The Outcrop Quizz.* Unwin Hyman, London.

Young, S., Ott, L. & Feigen, B. 1978. Some practical considerations in market segmentation. *Journal of Marketing Research*, **4**, 408.

23 Geology and the media

Anna Grayson

Summary

- Conservation is about public awareness, and the media has an important role to play in raising the public perception of environmental, including geological, issues.
- Whether through newspaper, radio, television or popular books, many opportunities exist to promote geology.
- However, the media is often underused and misunderstood by conservationists.
- In particular, a number of frequent mistakes are made when dealing with the media. These are discussed and the need for good working relations with the media and conservationists is emphasized.

There are a number of magazines about geology on the market: *Geology Today*, *Geoscientist*, *Earth Heritage* and the free-sheet *Down to Earth*. These are all excellent and professionally produced publications. They have very different styles and are written from different standpoints, but they have one thing in common: they are aimed at, and read by, people who already have a declared interest in geology. Readers of *Geoscientist* are likely to have considerable knowledge, whereas the majority of *Down to Earth* readers will be 'amateur'; either with knowledge, or with an express desire to attain knowledge. The majority of people do not have an interest in geology; they do not see its value or consider its relevance to everyday life. This paper is about getting the message about the importance of geology to people with no previous knowledge. To people who not only have no interest in the subject, but who do not want to be interested, and see no reason why they should be interested. There are plenty of them, around about 56 million in Britain alone. It illustrates why geology is important, and how we all can contribute to getting geology into the media.

Why should we bother?

Each and every human being lives on rock and off rock. We are surrounded by rock, particularly on our doorsteps, in its raw form as building stone, and the products of rock as building and manufacturing materials.

Geology is a fundamental science and as such it needs to be understood and appreciated. Without geology, and the products of rock, chemists would have nothing to analyse or synthesize, engineers would have nothing to engineer. Biological science would have neither flora nor fauna let alone any bio-diversity were it not for the wonderful story of life, told in the fossil record.

Geology draws on all the other main sciences; physics, chemistry, biology and astronomy and, in turn, feeds back into them. It is a wonderful medium for scientific education and justifiably an important part of the National Curriculum, although this is under threat.

Geology is a science to which amateurs can still make a valuable contribution. This was said about astronomy thirty years ago, the heyday of Patrick Moore, himself an amateur, but that barely remains true today. In palaeontology many of the most important finds have been made by amateurs, or amateurs turned professional with no formal academic training. Geology is a wonderfully healthly, convivial, outdoor activity, and a part of our heritage that is safe and accessible for every man, woman and child to enjoy.

Conservation of the countryside is politically correct and supported by most. Yet conservation is perceived as 'dicky birds, orchids and furry cuddly things'. The landscape, however, is made of rock. Animals, birds and flowers are dependent on rock just as we are. In the urban environment, we do not have many furry animals or flowers worthy of conservation, but we do have a great deal of rock, and it is under threat from developers, landfill and urban sprawl. There must surely be ample scope for earth science conservation and public awareness to be combined.

Perhaps most of all though, geology has such a cracking good story to tell; 'the last untold epic story' as Alan McKirdy of Scottish National Heritage put it to me during an interview for Radio Four. The story told in the rocks of our landscape is the story of how we come to be here.

How do we tell this story?

Beyond the professional magazines, the main stream media is the route to the untapped public. Here the importance of geology and the epic story it contains can be told using newspapers, radio, television and popular books; a process which is essential to raise public awarness of geology and promote its conservation. In this chapter, I will concentrate on the first three, at both local and national levels. Popular books are also important, but are not dealt with here.

There are two main ways to communicate a story to the press and media. One is through personal contact and the second is through the press release. Personal contact is often the best route, especially to journalists hungry for a story. This is, however, dependent on having the opportunity to make contacts and it is more likely, therefore, that most will have to rely on a press release. A press release is a brief printed statement, no more than two sides of A4, preferably one side, setting out the story. The first two sentences of a press release are vital - they must sell the story to the editor. He/she must be made to sit up and take notice, and must be able to see immediately why this particular story is relevant to his/her readers or audience.

In many organizations, the above is the job of a Press Officer, a rare beast in geological circles. The British Geological Survey have one in Ms Hilary Heason. Hilary nurtures contacts with the press (wining and dining us occasionally), and most importantly, sends out press releases. She does this selectively. There will be occasional stories of interest only to local journalists (e.g. a new regional map) with which she will not clutter the desks of national journalists. Sometimes a press release will result in wide coverage, sometimes no one picks it up, but geological stories rarely go away, and can often be 'sold' again at a later time. Perhaps most importantly, Hilary answers queries from the press promptly and politely and provides 'off the record' background information and briefing where necessary.

Who in the media could be approached?

Local papers and local radio are the most important. Local papers are always grateful to receive ideas for photo-stories. Most geological conservation stories will come under the headline of features, and it is the features editor who is likely to be your most useful contact.

Local Radio is a slightly different beast; they will be looking for two kinds of story: (1) the feature story, where they will send a reporter out to see you: for example, field trips with children involved always make good features; and (2) a live guest to come into the studio to chat.

On the national level, newspapers and radio, and very occasionally television, will be interested in big stories, but they do have to be really big and truly amazing stories in order to sell them successfully. If there is a major find in your area, or a really important part of our national earth heritage threatened, it is worth getting in touch either with a specialist journalist or with one of the science editors of the national press. On the national level you will be dealing with highly-trained professionals, so take care not to tell them their own jobs!

Television is always seen as the best medium and there is no doubt that it is a wonderful way of educating and informing the public. It is, however, extraordinarily difficult to get ideas and proposals accepted. People often say to me 'geology is a visual subject'; this is true, but that does not automatically make it a gift for television. Rocks do not move and we are talking moving pictures with television. Rocks often need an experienced eye to interpret, and a brain to highlight, what you are seeing. Unfortunately, the camera is not an experienced eye and it has no brain. Similarly, we see in three dimensions. Unfortunately, the camera tends to flatten a great deal that is obvious and gob-smacking to a geologist in the field. Television is expensive, requiring a minimum of £50 000 to make just one half-hour programme. Editors and channel controllers are not going to spend that sort of money unless they are absolutely convinced that the programme will sustain the attention of the viewer for the full duration, and I am talking about a viewer with no declared interest; a viewer holding a remote control, when Raquel from Coronation Street is about to be seduced by Jack Duckworth on the other channel. That is not to say the full-length feature programme on geology cannot be done.

Regional television news programmes, however, are well worth approaching with big stories or big events. Some television magazine programmes have picked up geological stories in the past. The programme

Country File picked up the limestone pavement story, while *Blue Peter* has occasionally run palaeontology stories.

National Radio, the medium in which I do most of my work, has proved to be a rich vein in recent years. *Rock Solid* still holds the record as Britain's longest running national media series on geology. *Science Now* also commissions geology pieces fairly regularly. Radio is a wonderful medium for geology since the subject involves the reconstruction of the past, and the pictures you can build up in someone's imagination can shine more brightly than the best visual image. Radio is better than any other medium for giving access to real people. It is the contributors, the 'real' scientists, that really make a radio programme. I am often asked what makes a good radio contributor, and it is worth stressing that it is personality rather than voice quality that is important; the best kind of broadcaster is the type of person who is quite relaxed about striking up a conversation in a pub with a stranger over a pint. Radio is a one-to-one conversation, relaxed and chatty with a microphone instead of a pint.

If it is so easy why have we failed in the past?

The geological community has frequently missed its opportunity to promote geology in the media by making the following mistakes.

1. It is easy to complain about the media. Many people have a wrong-headed perception of what the media do. Hopefully this paper will guide you, but if you are going to complain about the media coverage, make sure you have got your facts right and that you are complaining to the right person. This will prevent misunderstanding and help promote the real issue - the subject itself.
2. Directing the wrong story at the wrong people. The specialist scientific press is not interested in people dressing up in dinosaur costume. On the other hand, the popular press, the local press in particular, may love it. Similarly highly esoteric specialist stories are for the specialist magazines and journals, not for the press and media: the geomorphological evolution of sediment filled solution hollows in southern Algeria, or the occurrence of skeletal but euhedral high-Cr spinel in picritic minor intrusions in some God-forsaken minute Hebridean island, are, for examples, non-starters.

3. Failure to understand the law on copyright. It is amazing just how much confusion there is about this. There is no copyright on title but nicking a title may cause confusion and will not win friends and influence people! There is no copyright on ideas or on knowledge, but there most certainly is copyright on formatted material, i.e. knowledge and ideas presented in a particular way. Everyone is aware that there is copyright on published material. You cannot go about reproducing other people's published material without their permission, that is illegal. Yet I have come across a perception with one or two individuals that material broadcast by the BBC is in the public domain and is therefore up for grabs. This is most definitely not the case. You can quote the odd line for the purposes of review; this is termed 'fair dealing'. You can copy material for private study. BBC Education broadcasts may be recorded for teaching purpose, provided that those recordings have been made by, or for, the educational establishment concerned. If you want to use someone's copyright material, in another publication or in a public lecture or exhibition, all you have to do is write to them. You can use this as an opportunity to make more friends and to influence people; the copyright holder will be interested in what you are doing and may even be able to help you bear richer fruit. Sometimes a fee will be involved, almost always for pictures and video, but often as not, if it is a worthy cause, fees are waived. Most people are only too happy to see their work used for a wider audience. Even if you think that material might not be copyright, write and ask and inform individuals and organizations of your intentions. If in doubt, check it out. You should always credit someone's work or inspiration, even if fees and/or copyright are not involved. That is simple good manners and common sense.
4. Equal opportunities. Although geology is still male dominated, the majority of men are well behaved towards women, enjoy their company and value the contribution that they make towards the science and its public understanding. Sadly, there are some geologists whose attitudes are poor. Such attitudes are deeply ingrained, so much so that the perpetrators are often not conscious of the offence their behaviour can cause. Indeed, if you point it out, they are horrified! This can be a particular problem in the relations with the national broadcast media where so many of the commissioning editors and

programme makers are women - successful, well informed and educated women - who are in a position to refuse to tolerate bad behaviour and attitudes. All of the larger media organizations will have a policy on equal opportunities, and it is as well to bear this in mind.

5. The biggest mistake of all is a total failure to communicate effectively with journalists and the media. The media cannot even try to write or broadcast about geology unless you communicate with us, by letter, press release, or telephone. Worst of all are the numerous occasions when journalists have made enquiries and have met with no reply or with rudeness and hostility. This is unforgivable but all too common.

Conclusion

To finish I can do no better than to issue an impassioned appeal to those with influence in the world of geology to get their public relations act together. I asked Hilary Heason of the British Geological Survey what she thought the main function of public relations is: 'the object of public relations is to change four classic negative states into four positive ones in order to achieve the essential objective which is understanding: hostility into sympathy; prejudice into acceptance; apathy into interest; and ignorance into knowledge.' I rest my case and I look forward to hearing from you.

Acknowledgements

These are my own views and should not be taken as the official policy of the BBC.

Part Four

Creating an urban geological resource

This part explains how we should be proactive in creating a resource in areas where little or none exists, and in the better promotion of exciting resources.

24 The introduction of geology into the urban environment: principles and methods

Matthew R. Bennett & Peter Doyle

Summary

- The broad aim of earth heritage conservation can be best served by raising public profile of geology within urban areas.
- Conservation is about public awareness, and given that 80% of the population of the United Kingdom live in urban areas, it is here that we must promote the value of geology.
- This contribution sets out some of the principles and methods by which the geological component of our urban landscape can be increased and enhanced.

Conservation is about awareness and education; the more that people learn to value the natural environment, the greater the level of conservation achieved. Earth heritage conservation has long been the poor relation to wildlife conservation, primarily because of a lack of public awareness of the importance of geology within the environment. If we are to conserve our geological/geomorphological heritage, we need to promote public awareness and thereby exercise control through popular opinion on the planners, developers and decision makers who have the potential to threaten or conserve our earth heritage. To be conserved geology must be valued not simply by a few elite scientists, but by all. We need, therefore, a 'people-centred' conservation strategy (Carson & Harley 1996). This chapter is about increasing awareness of geology. The greatest potential for increased awareness is in the urban environment. Over 80% of the population in the United Kingdom live in urban areas: it is to these people that we must take the message of geology. To do this we must not only exploit the existing geological resource of building stones, parks and the like, but must also consider the creation of new resources where none exists. Every child in Britain is entitled to, and should obtain some education in geological principles (Hawley 1996). By establishing an urban teaching resource, generations of school children stand a better chance of being introduced to geology in such a way that its importance is appreciated and valued. One of the most important roles of urban geology is, therefore, to supply and support an educational resource for the hard pressed and often non-specialized teacher. After all, today's school children will have the power to conserve or destroy our geological heritage in the future. The aim of this paper is to set out the principles and methods available by which awareness of the earth science resource in urban areas may be increased and promoted.

Geology and urban areas

Geology has had a profound effect on the location and development of urban areas and is integrated into the very fabric of our cities and towns (Charsley 1996). Not only is it present in the building stones, but in most cases it controlled the location and commercial or industrial development of the town or city. This geological influence may not always be clear, yet in almost all cases urban growth will have been guided or shaped by geology at some stage in its development. If we are to raise the public profile of geology in urban areas, we must exploit this fundamental link between geology and urban development. In raising the public profile of geology the approach adopted will vary depending upon whether an urban area has any visible, naturally occurring geology. One can recognize three types of town or city on the basis of visible geological resources.

1. Those areas with a striking or visible geological resource. A classic example is Edinburgh: Arthur's Seat and Castle Rock are pre-eminent. Lyme Regis and Whitby are smaller towns whose economy has been influenced by geology both in the past and present; visitors still come to these towns specifically for their Jurassic fossils. Public awareness, amongst both the locals and visitors, of geology in all three examples is strong.

2. Those areas with some naturally outcropping geology. Towns with some geology include places like Frodsham, in Cheshire, where the high street is built directly upon hard, outcropping Triassic Sandstone; or Clitheroe in Yorkshire, which is surrounded by a ring of limestone quarries. In these towns public awareness of geology may be low and the local population may take it for granted.

3. Those areas with little or no exposed or visible geology. Most towns in the southeast of England fall into this category along with major cities such as Cardiff, Birmingham and London. Awareness of geology in these large urban conurbations is usually low and the link between a town and its geology may be completely lost.

The recognition of these three types of urban area is important in determining a strategy for the promotion of earth heritage awareness. In Type 1 and 2 urban areas, the focus is on the promotion of the existing geological resource in order to raise local awareness of the value of the natural geology around them. For Type 3 urban areas, with little or no visible geology, the focus must concentrate on how to create and introduce geology into the urban environment; and to exploit the artificial resources which exist. The different strategies required in these two broad types of town or city are discussed below.

Exploitation and promotion of existing resources

In urban areas which contain natural geology the focus for increasing public awareness lies with the local geological society, Regionally Important Geological/Geomorphological Sites (RIGS) group, museum and/or educational establishments (schools, colleges and universities). These groups have within their grasp the potential to influence local planners so that planning protection is granted to the naturally occurring geology through its inclusion in local structure plans. A two-way exchange is needed between the local geological society and the local officials or planners in order that the geology is recognized as part of the urban fabric. To achieve this local planners need to be introduced to geology and to its fundamental importance to the natural and commercial environment.

There are two basic principles which underlie the promotion of the natural resource in type one and two urban areas: (1) the natural geology forms part of the fabric of the town and city; and (2) the urban development and its success are intimately linked to geology. The importance of these principles is that they underlie the basis for the town, and as such, the promotion of the relevance of geology to the town is a relatively easy task.

It can be achieved by the tackling the following objectives.

- The geology/geomorphology of the town should be part of every signboard, leaflet or display even if it is only mentioned in passing. The link between the human and historical fabric of a town or place and its geology should be made at every opportunity. For example, if a town owes it location to a defensive position it should be clear from any display how this local geology gave rise to this defensive position.
- The link between the building stones, street furniture and vernacular architecture of a town and its local geology needs to be emphasized. This could simply take the form on modern buildings or stone street furniture of a small inscription giving the name of the rock and the location from which it was obtained.
- The educational potential of the local geology needs to be developed for schools by local geologists.
- Ideally the landscaped ground of parks and open spaces could be in harmony with and reflect local geology. For example, local stone should be used in rockeries and the natural structures common in the areas should be mimicked.

In these ways it is possible to raise the public profile of geology in urban areas with good natural geological exposures. The British Geological Survey is producing posters showing these relationships for British towns. In other cases the local geological society is vital. The success of geological conservation within Dudley owes much to the impact of the local geological society (Cutler 1996), and illustrates what can be achieved by raising the public profile of geology and establishing its link with the historical and cultural development of the area.

Creation of new geological resources

In Type 3 urban areas with little or no exposed geology, there are four basic resource areas: temporary exposures; museums; building stones; and, the creation of artificial landscapes or exposures.

Temporary exposures have the potential to allow one to observe real geology (Worton 1996), but are often inaccessible to all except the engineer or semi-professional geologist. Museums have a vital role to play but are poor in providing context for geology. They are also accessible only to the museum-going public. Building stones and the creation of artificial resources remain the most accessible methods of enhancing public awareness of geology. These are discussed below.

Recognition and promotion of the artificial resource

Within this volume we have seen the nature and extent of the geological resource which exists in urban areas. The potential of stone in cemeteries and buildings has been demonstrated (Mason 1996; Robinson 1996*a*). There is also a valuable resource within parks and museums (Doyle *et al.* 1996; Larwood & Page 1996), while the temporary excavations cut during development work also have potential (Worton 1996). In areas with little or no natural geology we need to exploit and maximise the potential of this resource.

This principle is well illustrated by the use of building stone. An increasingly diverse range of rock types is now used as facing stones on new buildings. For example, within 500 m of Liverpool Street Station in London there are over 50 different rock types used as facing stone on modern office blocks (Fig. 24.1). The huge potential for increased awareness of geology is lost because for all except the expert petrologist the rock types and their origins remains a mystery. If, however, small discrete inscriptions were put at eye level stating the rocks' name, age and source it would open up this resource to both local teachers and to the passing pedestrian. This type of information is often held by the architects or builders, and would cost very little to inscribe on the exterior of the building. This has been done to good effect outside Euston Station

Fig. 24.1. Photograph of a square in front of modern office blocks behind Liverpool Street Station, London. Note the rectangular blocks of stone laid for seating.

Fig. 24.2. Photographs of the stone benches and inscriptions of characteristic British lithologies in front of Euston Station, London.

in London where there are four stone benches each made of a characteristic British lithology (Cornish granite, St Bees Sandstone, Portland Stone and Cumbrian 'slate' tuff). On the plinth of each the rocks name and age is carefully inscribed and demonstrates geological age and diversity the different character of igneous and sedimentary rocks (Fig. 24.2).

Alternatively a business or shop proud of its new building could simply produce an A4 flyer with a labelled sketch of the building showing the different stones and their names and country or region of origin. If this was distributed to local schools it would be of immense value to them (Hawley 1996). The development of leaflets and trail guides to the petrology of ones local high street or cemetery have a very important role to play in exploit the existing resource (Robinson 1996*b*). A good example is that produced by BACMI (1993).

In promoting the potential geological resource of urban areas a number of simple objectives can be recognized. The owner/occupier of a site or building with geological potential could be assisted in developing its potential value to local schools. Local geologists and societies need to publicize the nature of the geological resource of their urban environment for local schools; ideally this should be done in partnership with the schools and owner/occupier of the site involved. The geological community as a whole needs to encourage the users of stone to label the stones used wherever possible. If this could be established as 'good practice' it could have a dramatic effect on the potential of the artificial geological resource of urban areas.

Clearly the local geological society, museum or amateur geologist has an essential role to play in raising awareness of the artificial geological resource within our urban areas. Ideally local geological societies should be encouraged to prioritize the development of their urban geology, represented by the building stones around them.

Creating a new geological resource

In Type 3 urban areas with little of no natural geology it is possible to create a new geological resource. In most urban areas of Britain there is constant turnover and development of our urban areas, as new areas are developed or old ones revitalized. There is a potential within this process for the creation of new geological resources. Worton (1996) has already described how the involvement of local geologists at the planning stage of projects has the potential to achieve geological gain in areas with an existing geological resource. The key to success is to get involved at the initial planning stage. In the same way geological involvement in the planning process of any development work has the potential to create new artificial geological resources, particularly in areas with landscaped ground.

Fig. 24.3. Photograph of Richard Harris's sculpture in front of the Royal Festival Hall on London's South Bank. The sculpture is constructed from narrow concrete flags, like those commonly available in garden centres.

Many development projects involve landscaping and it is this which has the greatest potential for the creation of new geological resources. There are two principle areas in which an artificial resource can be created: the creation of geological illustrations, and the creation of rock-works.

Geological illustrations

Within landscaped areas there is a wide range of different types of geological illustration that can be constructed. The stone benches cut from different lithologies in front of Euston Station in London, provide a good example of what can be done (Fig. 24.2). The inclusion of inscriptions of the name, origin and age of facing stone used in buildings would be a simple task, or alternatively inscribed sample blocks of the stone used could be worked into the street front or into a landscaped or sculptured area. Figure 24.1 shows a modern square between office blocks behind Liverpool Street Station in London, where polished blocks of stone are scattered throughout the square as seating. If these blocks were inscribed with the name, origin and age of the stone or if a range of different blocks were used and labelled, this would have much greater value as an educational resource and in increasing general awareness of geology.

Other geological illustrations can be constructed. Many urban parks or landscaped areas contain sculptures and decorative walls. Figure 24.3 shows a sculptured wall in front of the Royal Festival Hall on London's South Bank which illustrates the geological potential of such structures. This sculptured wall and the flight of steps within it, by Richard Harris, are constructed from concrete flags and perhaps unconsciously illustrate how the outcrop of dipping strata varies around a valley, the valley being represent by the central flight of steps (Fig. 24.3). Such a sculpture could be used to teach students how to measure the angle of dipping strata and demonstrates important geometric principles. Figure 24.4 illustrates how a series of simple geological structures could be created in walls out of concrete flags. Such geological illustrations could be used to teach such principles as superposition as well as providing an opportunity to measure features such as dips and fault throw. They can be made from a variety of different material: bricks, concrete flag stones, stone block work, concrete. For example, Figure 24.5 shows simple folds created from blocks of Jurassic sandstone in Valley Gardens Park in Scarborough. Such illustrations provide textbook examples combined with the unique hands-on experience of geology field work. This type of illustration does not need to be made from natural stone to be effective.

The construction of problem walls is analogous to the development of artificial climbing walls in urban areas. Problem walls may consist, for example, of walls constructed around blocks of stone of different

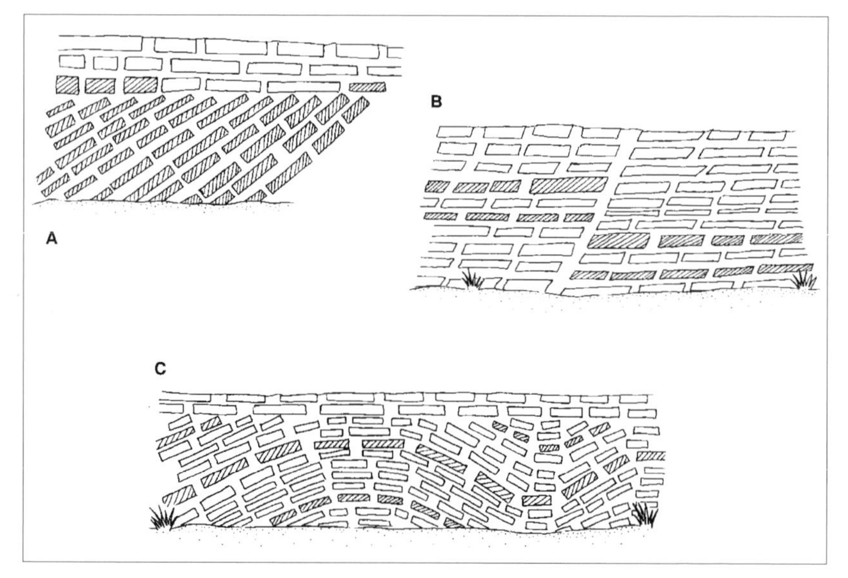

Fig. 24.4. Cartoons of simple geological illustrations which could be built into walls. **(a)** Angular unconformity; **(b)** fault; and **(c)** folding.

Fig. 24.5. A geological illustration of simple folds constructed from Jurassic sandstone blocks in Valley Gardens Park, Scarborough, North Yorkshire.

lithology, which would allow such education exercises as the 'wall game' to be played (Robinson 1996*b*) (Fig. 24.6a). The different lithologies could be obtained from the stone masons' waste skip or from the suppliers of architectural facing stone. Alternatively walls could contain different fossils, perhaps in evolutionary order. More ambitious walls could contain way-up structures, manufactured on site. For example, different pebbles pushed into a concrete surface could represent graded bedding, and be placed within the wall along with blocks of natural sandstone with such containing structures such as truncated cross bedding (Fig. 24.6b). With a little imagination some very interesting problem walls could be generated. They do not necessarily require expensive materials, simply using those which can be obtained locally and are often regarded as waste. The potential is considerable if landscape planners and architects could be made aware of the educational potential of very simple structures. Many new schools now have wildlife ponds and nature areas within their grounds, so why not also include geological walls?

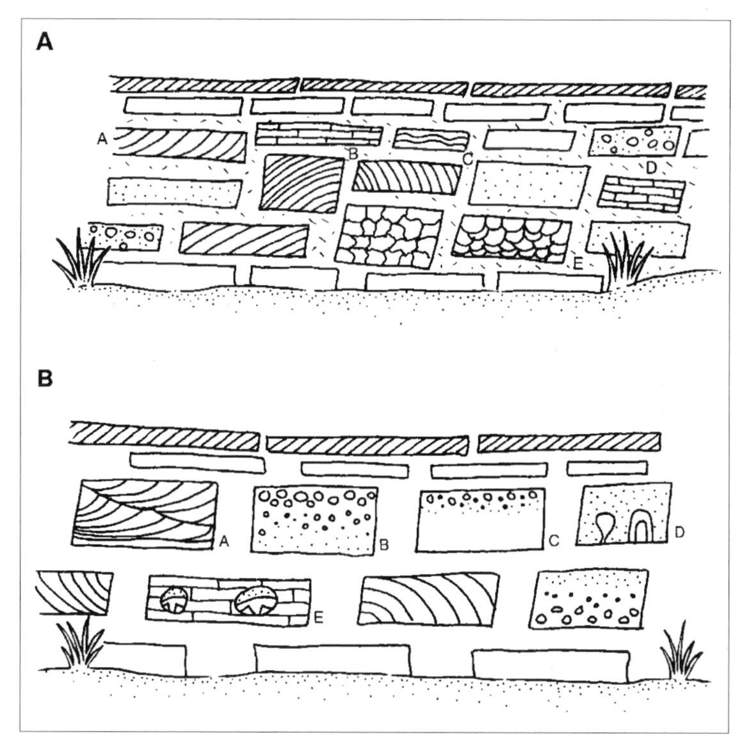

Fig. 24.6. Cartoons of geological 'problem' walls. **(a)** Wall made up of various different lithologies, on which the 'Wall Game' could be played (see Robinson 1996*b*). **(b)** Wall made up of blocks with different way-up criteria within them.

Rock-works

The creation of rock-works has a much wider potential than just education, since it provides a means of introducing an appreciation of geology at a subliminal level to a wider cross-section of the general public.

Victorian landscape gardens were particularly innovative in the construction of rockeries and artificial rock-works (Doyle & Robinson 1993, 1995; Robinson 1994; Doyle *et al.* 1996). The more sophisticated rock-works of this time varied from geological models or reconstructions using the correct materials such as those at Crystal Palace to completely imaginary rock-works, such as those at Battersea Park, constructed from artificial stone (Doyle *et al.* 1996). What characterizes these Victorian rock-works is the level of geological observation which went into their construction. Immense trouble was taken to replicate the detail of natural rock exposures (Fig. 24.7). It is this attention to geological principles which makes them so successful in portraying geology, to such an extent that geologists have frequently mistaken them for natural exposures (Doyle *et al.* 1996). In contrast, modern rockeries rarely attain such a natural perspective and the rock within them frequently looks artificial (Fig. 24.8). Today landscape gardeners rarely appear to apply the simple geological

Fig. 24.7. Photograph of a Pulhamite crag at Battersea Park, London.

Fig. 24.8. Photograph of a modern rockery.

principles which were followed by their Victorian predecessors. Modern attempts at rock-works look artificial and, instead of mimicking the natural interaction of rocks and plants, simply provide a sculptured back drop for the plants. As a consequence many modern rockeries have little geological potential and do not foster an appreciation of geology (Fig. 24.8).

The Victorian enthusiasm for natural rock-works stems from the popularity of geology and natural history at this time, a time when rocks and fossils were of national interest and debate. This interest in rock gardens and rock-works is typified by the publication of a volume on rock gardens and alpine plants by Robinson (1870) which is still the seminal work on the subject. This volume sets out the principles behind the construction of realistic rock gardens on the basis of geological principles and appears to have formed the basis for almost every subsequent publication on rock gardening since (e.g. Meredith 1910; Thomas 1917; Farrer 1919; Anderson 1959; Foster 1982). Despite this, most contemporary rock garden manuals contain little sound geological guidance, and this may help explain why so many modern rockeries look unnatural (Bird 1990). If rock gardens on modern developments could be constructed more frequently along geological lines, it would not only enhance their aesthetic impact but would also help foster a greater appreciation of the aesthetic potential and therefore value of rock. The methods by which a geologically convincing rock garden can be constructed are presented in the appendix to this chapter.

Unfortunately, in recent years there has been a reaction within the conservation community against the construction of rockeries because of problems associated with the exploitation of waterworn limestone and other naturally weathered rock. Limestone pavement with its delicately sculptured morphology has long been prized by the horticulturalist. This demand has accelerated the rate of destruction of this finite and unique geological resource within the British Isles (Goldie 1994). In recent years some campaigners against the use of limestone pavement have tried to foster a false sense of fear against the use of any rock in gardens or landscaped ground (Carson 1995). However, as pointed out by Bennett *et al.* (1995), although the use of limestone pavement is wrong and should stop, the use of other types of rock in gardens and landscaped ground is beneficial to geological conservation. The use of natural stone in gardens has few conservation implications provided that limestone pavement and other weathered stone is untouched. Natural stone is a by-product of commercial quarries, driven by the needs of the building industry and not by landscape gardeners. Waste stone is always produced as a by-product of quarrying and is just as suitable for use in rockeries and landscaped ground as limestone pavement. In urban areas stone may also be recycled from demolition sites.

Where natural stone is not available artificial stones can be used. The techniques used by Pulham can be recreated by using 'shotcrete' or 'guncrete' concreting techniques over a brick and rubble base to produce rock-works. Equally excellent artificial rocks can be bought commercially. They are produced primarily for the theatrical, film and display industry out of fibreglass and polystyrene. Home made rocks can also be constructed using cement and grit aggregate from polythene lined moulds shaped from soil or sand. Although some earth scientists have argued against the use of artificial and imported stone (Carson 1995) this is short sighted and is a view which does not appreciate the value of rock gardens in promoting awareness of the relationship of geology to the landscape and the plants within it.

The rock garden is an important way of introducing geology into the urban landscape. It provides a way of illustrating to a large audience the natural aesthetic beauty of rock and its intimate association with plant habitats. At a subliminal level it has the potential to interest a wide audience in geology, and as such is of considerable importance to urban geology. As geologists we should encourage the use of stone, artificial or natural, in gardens and landscaped areas. Its use raises the profile of geology and encourages people to appreciate and value it, in the same way as they appreciate and value plants.

The principles and objectives of creating new geological resources in areas with little or no natural resources are summarized below.

- Local geologists need to get involved early in the planning process of developments with a potential geological interest as well as those in which geological illustrations, educational resources and rock-works can be built into the project. This will only happen if the geological input occurs early enough before the plans are finalized and the contracts let (Worton 1996). At present local geologists tend only to get involved where a RIGS or Site of Special Scientific Interest (SSSI) exists. We should, in fact, take a greater interest in the whole planning process, whatever the nature of the development. There is potential for geological gain in every development whether it be a housing estate, factory or road.
- The geological community as a whole needs to demonstrate the potential of geological illustrations and rock-works to planners, architects and landscape gardeners.
- Any new resource created is of little value unless it is used by local schools, appreciated by the local community or valued by the site owner/occupier.

Finally, as geologists, we tend to value and appreciate only natural rock exposures. Yet if we are to raise the public profile of the earth sciences in urban areas, particularly those with no naturally occurring geology, we must become involved with the creation of new geological features and resources within the urban environment.

Conclusion

One of the most important parts of any earth heritage conservation strategy is the battle for public awareness. If people understand and appreciate the value of geology, its conservation is more assured. Since the majority live in urban areas we need to raise the profile of geology in these areas. In towns and cities with natural geological exposures this task is easier and is one which centres upon publicising the geology and integrating it into the community. In towns and cities without any natural exposures we need to exploit the geological resource provided by such things as building stones, parks, museums and artificial rock-works. We must also consider the importance of creating new geological resources as a consequence of urban renewal and development. To do this we need to encourage and foster urban architects, developers and planners to think about introducing geology into developments and to consider its potential benefit, particularly for education, to the urban community.

The principles and ideas presented in this paper range from the very simple to the more complex, but the central principles remains the same. We need: (1) to exploit the existing geological potential of our urban environment; and (2) to educate architects, developers and planners in the simple ways that they can introduce geology into our towns and cities. In this way we should be able to help foster greater interest in, and support for, earth heritage conservation.

References

Anderson, E. B. 1959. *Rock gardens.* Penguin, Harmondsworth.

BACMI1993. *Rocks around you: Earth Science for Key Stage 3.* Hobsons Scientific and the British Aggregate Construction Material Industries (BACMI), London.

Bennett, A. F., Bennett, M. R. & Doyle, P. 1995. Paving the way for conservation? *Geology Today*, 98-100.

Bird, R. 1990. *A guide to rock gardens.* Christopher Helm, London

Carson, G. 1995. Letter to the editor. *The Guardian*, 21 January.

---- & Harley, M. 1996. Shifting the focus: a framework for community participation in earth heritage conservation in urban area. *This volume.*

Charsley, T. J. 1996. Urban geology: mapping it out. *This volume.*

Cutler, A. 1996. The role of the regional geological society in urban geological conservation. *This volume.*

Doyle, P. & Robinson, J. E. 1993. The Victorian 'geological illustrations' of Crystal Palace Park. *Proceedings of the Geologists' Association*, **104**, 181-194.

---- & ---- 1995. Report of a field meeting to Crystal Palace Park and West Norwood Cemetery, 11 December 1993. *Proceedings of the Geologists' Association,* **106**, 71-78.

----, Bennett, M. R. & Robinson, E. 1996. Creating urban geology: a record of Victorian innovation in park design. *This volume.*

Farrer, R. 1919. *The English rock garden.* T. C. E. C. Jack Ltd, London.

Foster, H. L. 1982. *Rock gardening*. Timber Press, Portland Oregon.

Goldie, H. S. 1994. Protection of limestone pavement in the British Isles. *In*: O'Halloran, D., Green, C., Harley, M., Stanley, M. & Knill, J. (eds) *Geological and Landscape Conservation.* Geological Society, London, 215-220.

Hawley, D. 1996. Urban geology and the National Curriculum. *This volume.*

Larwood, J. G. & Page, K. 1996. Museums: a focus for urban geology and geological site conservation. *This volume.*

Mason, R. 1996. Heathen, xenoliths and enclaves: kerbstone petrolgy in Kentish Town, London. *This volume.*

Meredith, L. B. 1910. *Rock gardens: how to make and maintain them.* Williams and Norgate, London.

Robinson, J. E. 1994. The mystery of Pulhamite and an outcrop in Battersea Park. *Proceedings of the Geologists' Association,* **105**, 141-143.

---- 1996*a*. 'The paths of glory. . .' *This volume.*

---- 1996*b*. A new version of the wall game in Battersea Park. *This volume.*

Robinson, W. 1870. *Alpine flowers. for English gardens.* John Murray, London.

Thomas, H. H. . 1917. *Rockeries: how to make and plant them.* Cassell & Co, London.

Worton, G. J. 1996. Digging up your doorstep: engineers and their excavations. *This volume.*

Appendix: The construction of rock-works and rockeries

Traditionally rockeries or rock-works are classified into: moraines or screes, which consist of a loose collection of blocks; walls, in which plants emerge from soil filled crevasses between each block; and, natural outcrops. It is the last type of rockery which has the greatest potential benefit for geology. In construction of such rockeries one must consider the choice of materials and the placement and design of the actual outcrop.

Materials

With regard to the choice of materials the following points are of note: never use limestone pavement, weathered gritstone, weathered granite or stone taken from dry stone walls; wherever possible use local rock types; where natural stone is not available artificial stone should be used. Artificial stone can be bought commercially or made from fine grained concrete plastered over buttresses of rock formed from block work, or made in earth or sand moulds from coarse grained cement.

Construction

The best guide to the construction of rock-works can be obtained by evaluating the reasons why the large Victorian reconstructions at locations such as Crystal Palace and Battersea Park appear so convincing, despite the fact that they made from artificial material. These rock-works are convincing because: (1) there is continuity of the bedding or horizontal joint surfaces within and between rock outcrops - there is a constant dip; (2) there is continuity to the vertical joints sets present; (3) the individual blocks of rock are in proportion with the scale of the slope - were the bank is large the individual blocks are large; (4) the individual blocks of rock emerge from the hill and appear to form the slope or hillside; and (5) attention to geological detail.

The following principles should be followed in the construction of rock-works.

- The amount of rock used should be in proportion with the area of the rockery. As illustrated in Figure 24.9 if one uses too few rocks in a given areas they appear like a boulder field while too many rocks within an area may appear too much like a wall. More importantly

Fig. 24.9. Diagram to show how the amount of rock used in a rockery may affect its visual appearance. **(a)** Too few rocks; **(b)** too many rocks.

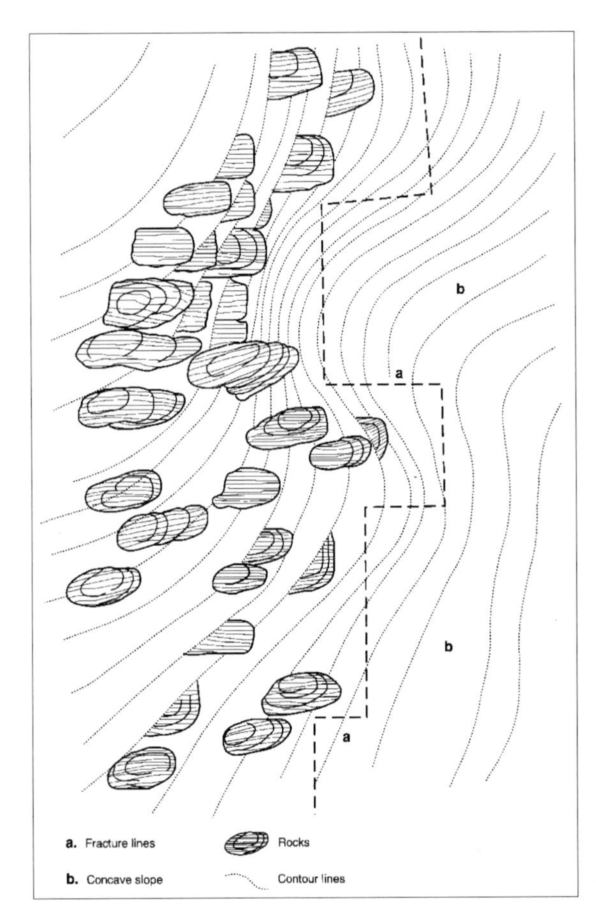

Fig. 24.10. Contour map showing a vertical plan of a rockery.

the amount of visible rock varies across a slope. The level of natural outcrops varies with the slope from which they emerge (Fig. 24.10). The rocks must appear as if they form the slope or hillside from which they emerge

- There must be continuity in the bedding or horizontal joint surfaces across and between the rock outcrops. Sedimentary rocks are laid down in layers. Blocks within rockeries should also be laid down in layers (Fig. 24.11). The bottom surface of each block should be either be horizontal, or, if a dipping outcrop is being constructed, they should all dip in the same direction and by the same amount. The base of

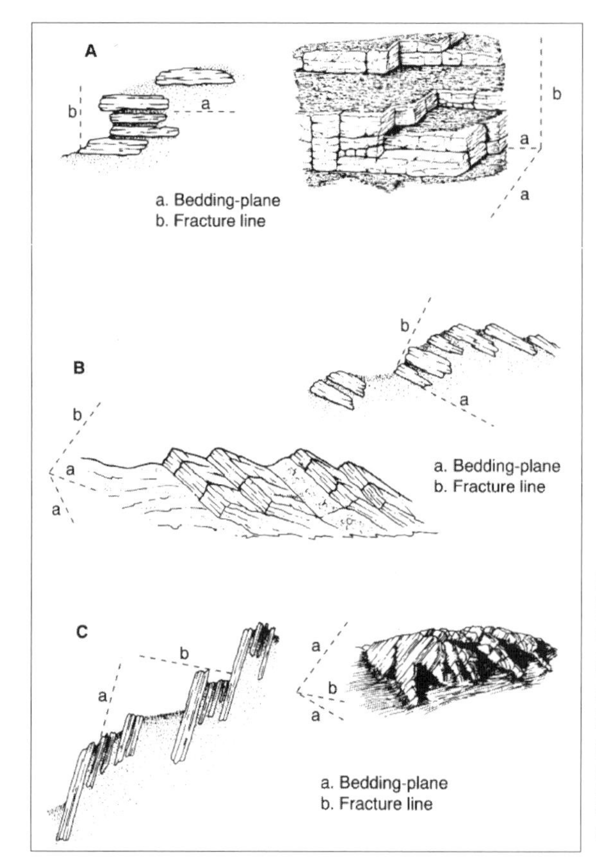

Fig. 24.11. Diagrams to illustrate how the vertical joint sets should be orientated at right angles to the chosen bedding surface. **(a)**Horizontal bedding; **(b)** gently dipping bedding; **(c)** steeply dipping bedding.

blocks should line up along contour lines drawn on the slope (Fig. 24.10). The greater the continuity of these layers across the rockery or garden the more effective will be the effect. One of the most impressive features of Battersea Park Pulhamite rock-works is the continuity of the bedding surfaces, not just within one outcrop but from one outcrop to the next (Fig. 24.7). It is probably this single factor which makes the reconstructions so convincing. Modern additions to the Battersea rocks look out of place because they dip at discordant angles, and are out of place with the overall continuity of the bedding. Even igneous rocks in the natural environment could contain some form of structure or banding which needs to be orientated in a consistent fashion.

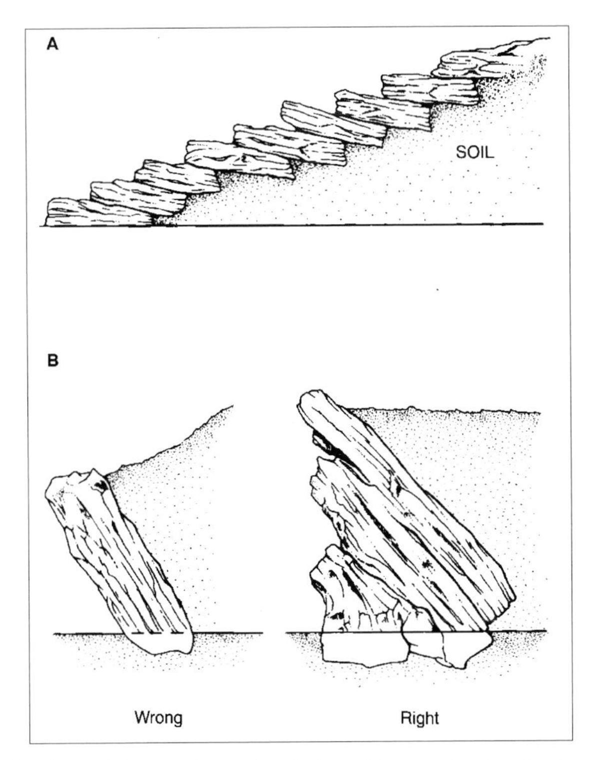

Fig. 24.12. Rocks within a rockery should be laid so as to give the impression of dipping strata. **(a)** A series of blocks laid down to represent a gently inclined dipping surface. Note the continuity in the grain within each block which parallels the imaginary bedding surface. **(b)** A series of blocks laid down along a steeply inclined imaginary bedding surface; this also indicates the need for firm foundations in the placement of each block.

- There must be continuity in the vertical joint sets. Vertical joints are represented by the edges of individual blocks. These need to be orientated at right angles to the chosen bedding surface as shown in Fig. 24.12.
- Not only does the base of each block need to follow an imaginary bedding surface, but it needs to parallel any horizontal structure (e.g. laminations, graded bedding, foliations, cleavage, jointing, etc.) present within each block (Fig. 24.13d). To the non-geologist this can be regarded as the 'grain' within the rock. This may mean that the face

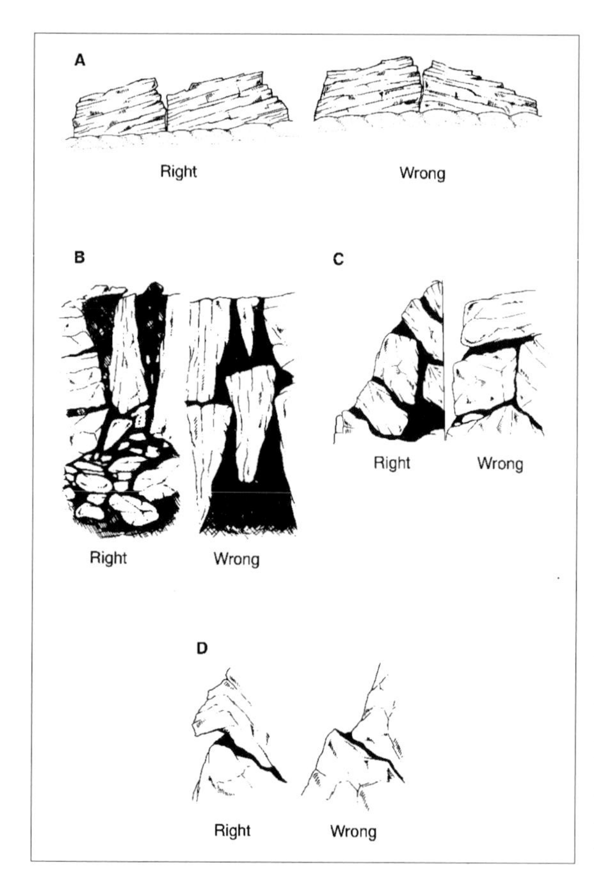

Fig. 24.13 (a-d). Good and bad practice in the construction of rock fissures between stone blocks in a rockery. **(a)** illustrates the importance of lining up the structural grain within each block.

of the block of stone chosen as the bottom surface may not be the largest or flattest face.

- Individual stones must be firmly placed within the soil and appear as if they emerge from it rather than having been placed in it (Fig. 24.11). This requires each stone or block to be carefully dug into the slope and placed in a compacted pocket of pounded earth.
- If fissures between blocks are to remain soil filled and successfully support plants they need to: incline upwards, or else the soil will simply wash from them (Fig. 24.13a); narrow downwards (Fig. 24.13b); and not be overhung by or over shadowed by other blocks (Fig. 24.13c).

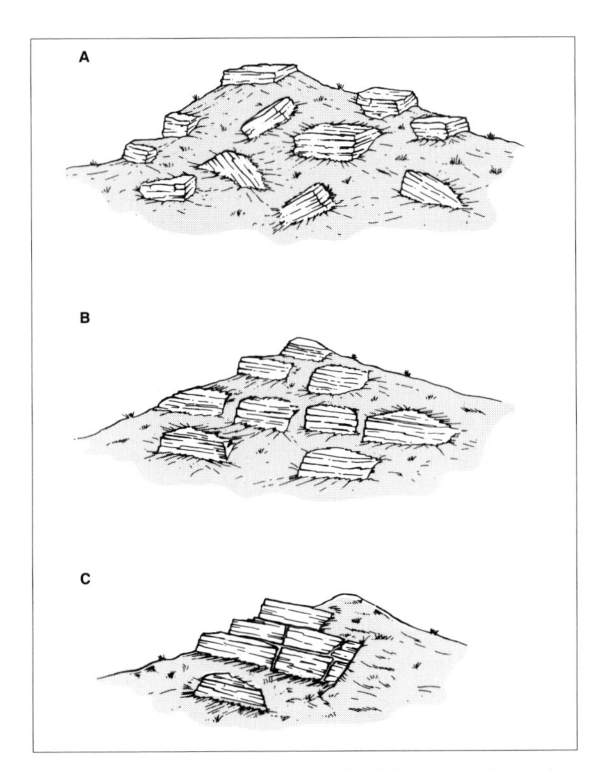

Fig. 24.14. Three sketches of a rockery. **(a)** The rocks have been placed at random, with no regard for maintaining any continuity of bedding, jointing or grain within the blocks. **(b)** The rocks have been reorientated so that there is a continuity in the bedding, jointing and grain of the blocks. **(c)** The blocks have been clustered to form a natural outcrop

The application of these principles in the construction of a rockery is illustrated in Fig. 24.14. The rock garden in Fig. 24.14a does not look natural because the rocks are organized in an irregular pattern, with no continuity in the bedding, joints or grain of the individual blocks. The rockery in Fig. 24.14b is more natural since the blocks are placed with respect to a uniform bedding surface and the grain of individual blocks has been matched up. However, the blocks still look as if they have been placed into the hillside instead of emerging from it. This can be rectified if the blocks are clustered to form a natural outcrop as shown in Fig. 24.14c.

Index

References in *italics* are to Figures and Tables